S. Hrg. 113–653

ASSESSING THE RISKS, IMPACTS, AND SOLUTIONS FOR SPACE THREATS

HEARING

BEFORE THE

SUBCOMMITTEE ON SCIENCE AND SPACE

OF THE

COMMITTEE ON COMMERCE, SCIENCE, AND TRANSPORTATION UNITED STATES SENATE

ONE HUNDRED THIRTEENTH CONGRESS

FIRST SESSION

MARCH 20, 2013

Printed for the use of the Committee on Commerce, Science, and Transportation

U.S. GOVERNMENT PUBLISHING OFFICE

94–799 PDF   WASHINGTON : 2015

CONTENTS

ASSESSING THE RISKS, IMPACTS, AND SOLUTIONS FOR SPACE THREATS

WEDNESDAY, MARCH 20, 2013

U.S. SENATE,
SUBCOMMITTEE ON SCIENCE AND SPACE,
COMMITTEE ON COMMERCE, SCIENCE, AND TRANSPORTATION,
Washington, DC.

The Subcommittee met, pursuant to notice, at 10:04 a.m. in room SR–253, Russell Senate Office Building, Hon. Bill Nelson, presiding.

OPENING STATEMENT OF HON. BILL NELSON,
U.S. SENATOR FROM FLORIDA

Senator NELSON. Good morning. We are delighted to have this meeting of our Science and Space Subcommittee in the new Congress.

NASA and the space programs have been in the news a lot in the past year. Some really impressive feats. And we are going to be talking about some of those, from the Rover on Mars to the berthing of the SpaceX capsule at the International Space Station.

I am delighted to have my colleague, Senator Cruz from Texas, as our Ranking Member. It seems like that Texas and Florida have some interest in the space program. And I am looking forward to his leadership. And I would ask for his opening statement.

STATEMENT OF HON. TED CRUZ,
U.S. SENATOR FROM TEXAS

Senator CRUZ. Well, thank you, Mr. Chairman. Let me echo those sentiments and say how much I am looking forward to working with you on this subcommittee.

Space flight and our capacity to maintain world-leading advantage in space flight is a critical priority for the nation and certainly a critical priority both for the State of Texas and the State of Florida. And so I am eager for our collective journey to ensure that NASA and all of the related programs have sufficient resources, sufficient priority to do what needs to be done.

And I appreciate all of the witnesses coming here today to address these important topics and also to begin the process of what I hope this subcommittee will do over the coming years, which is continuing to make the case to the American people about the importance of these programs, about the benefits that they produce for the private sector, that they produce for men and women across this country, and, at the same time, looking for ways to improve those benefits, to expand the cooperation that we presently have between NASA and the private sector, and to look for ways to even further increase positive benefits that are realized in everyone's day-to-day life.

And so I am eager and looking forward to working together.

Senator NELSON. Thank you, Senator.

In the interest of time, we are going to compress things today because there have been three roll call votes called at 11:15. So I would like to see if we can get the bulk of the hearing in the time before the votes take place so that we can be mindful of your time, because those votes will stretch out over some period of time.

And we are going to get into some interesting stuff here today on space debris and also asteroids possibly hitting the Earth. So we want to have time to cover this. Let me suggest to each of you, keep your comments to 5 minutes so that we will have a chance to get in depth in some of the questions.

We have Dr. Jim Green, Director of the Planetary Science Division in NASA's Science Mission Directorate.

We have former astronaut Dr. Ed Lu—two shuttle flights and a 6-day stay on the International Space Station. He is now the Chairman and CEO of the B612 Foundation, and he is going to talk to us about his foundation's Sentinel, which is to track near-Earth objects.

And then Mr. Richard DalBello, Vice President of Government Affairs for Intelsat, who is going to speak about the economic role of satellites and the commercial and security implications from the space threats.

And then Dr. Joan Johnson-Freese, Professor of National Security Affairs at the U.S. Naval War College. She is going to talk about the role of space in our daily lives and how space threats can threaten our national security.

I will put my formal statement in the record.

[The prepared statement of Senator Nelson follows:]

PREPARED STATEMENT OF HON. BILL NELSON, U.S. SENATOR FROM FLORIDA

Good morning. Thank you all for being here today for the first Science and Space Subcommittee hearing of the new Congress. NASA and the U.S. space program have been in the news a lot over the past year for some impressive feats of technology and engineering, from landing a rover on Mars to SpaceX berthing for the first time with the International Space Station.

But about a month ago, there was some news that was more scary than exciting. That's when a meteor exploded over Russia with more energy than 20 atomic bombs, shattering glass and injuring over 1,000 people along the way. And, that same day, an asteroid passed closer to Earth than we've seen in a while. The days' newspapers read like sci-fi movie scripts, but all the content was real. The threat from these near-Earth objects, as well as threats from space weather, debris, and more, deserves a closer look from this committee.

What have NASA and private space efforts done to increase our awareness of these space threats? And, what is being done to protect us and the systems we rely on from these threats? I'm looking forward to hearing more about that from the experts here today.

Like others who have traveled into space, I myself am no stranger to space threats. After we landed from my shuttle mission in 1986, NASA's post-landing debris damage assessment found that Columbia received several debris impacts.

The orbiter took a major hit on the Orbital Maneuvering System—the rocket engine used to perform orbit adjustments—and two craters were found around the right side window, the window that NASA's Administrator Charlie Bolden used when piloting *Columbia*.

It is also prudent to point out that the International Space Station had to do three collision avoidance maneuvers last year to dodge debris from both the Chinese anti-satellite test in 2007 and the Iridium—Cosmos collision of 2009.

But before I introduce our witnesses, I want to take a minute to remind everyone about the important role space plays in our lives, from the GPS navigation systems some of you used to find your way here today to the communication satellites that are allowing our remote viewers to watch this hearing. We live in a world that relies on systems in space and on the ground that are susceptible to space threats and we need to protect them just like we need to protect our planet itself.

Without further ado, it is my pleasure to welcome:

Dr. Jim Green, Director of the Planetary Science Division in NASA's Science Mission Directorate, who will provide an overview of NASA's work in these areas;

Former NASA astronaut Dr. Ed Lu. After two shuttle flights and a six month stay on the International Space Station, Dr. Lu is now the Chairman and CEO of the B612 Foundation and will talk to us about the foundation's Sentinel mission to find and track near-Earth objects;

Dr. Richard DalBello, Vice President of Legal and Government Affairs for Intelsat General Corporation, who is going to speak about the economic role of satellites and the commercial and security implications from space threats; and

Dr. Joan Johnson-Freese, Professor of National Security Affairs at the U.S. Naval War College. She is here to talk to us about the role of space in our daily lives and how space threats can disrupt national security.

I thank you all for being here today and look forward to your testimony.

Senator NELSON. Your written testimony will be inserted in the record, and if you would just give us a quick summary so we can get into the questions, please.

Senator CRUZ. And, Mr. Chairman, I will confess, given the topic today, I was disappointed that Bruce Willis was not available to be a fifth witness on the panel.

[Laughter.]

Senator NELSON. We might get a trailer from "Armageddon" and show that.

[Laughter.]

Senator NELSON. Dr. Green?

STATEMENT OF DR. JAMES GREEN, DIRECTOR, PLANETARY SCIENCE DIVISION, SCIENCE MISSION DIRECTORATE, NATIONAL AERONAUTICS AND SPACE ADMINISTRATION

Dr. GREEN. Mr. Chairman and members of the Subcommittee, I am pleased to have the opportunity to update the Committee on NASA's programs and our approach to addressing the risks, impacts, and solutions for space threats.

One space threat is Near-Earth Objects, or NEOs. NEOs are asteroids and comets that enter the near-Earth space. They are primitive leftover building blocks of the Solar System, making them also compelling objects for scientific study.

Today we do not have a complete inventory of all the possible impactors. NASA was tasked by Congress in 1998 to catalog 90 percent of all the large NEOs within 10 years. The large NEOs are those that are 1 kilometer or more in size. A large NEO would cause a global catastrophe if one struck the Earth. NASA now is cataloging up to an estimated 95 percent of all the NEOs over 1 kilometer in size. That said, none of these known large NEOs pose any threat of impact to the Earth within the next 100 years. It is a situation we are constantly monitoring.

In 2005, Congress directed NASA to expand the survey to detect, track, and catalog NEOs equal to or greater than 140 meters in diameter. Congress set a goal for this program to be 90 percent com-

4

pleted by 2020. For this expanded survey, NASA's NEO program currently has three survey teams that operate five ground-based telescopes and are providing the nonprofit foundation B612 with technical assistance and operational support through a Space Act Agreement for their space-based survey telescope that we will hear about today. NASA continues to make daily progress on this goal.

Extreme space weather is another threat being studied by NASA. Space weather refers to the conditions on the Sun and in the solar wind, and in the near-Earth environment.

Our ability to understand the Sun-Earth system is of growing importance to our Nation's economy and national security. The electric power industry is susceptible to geomagnetically induced currents, which can overload unprotected power grids and result in widespread power outages. In the spacecraft industry, intense geomagnetic storms have the capacity to disrupt normal operations, such as satellite communication. And, of course, they pose risks to our astronauts in space. In addition, space weather can cause irregularities in the signals from our very important Global Positioning System.

The National Oceanic and Atmospheric Administration, or NOAA, is the official source for space weather predictions for the Nation. Several of NASA's research satellites have become an essential part of our Nation's space weather prediction system, providing very important data for determining the space weather conditions. One such mission is the Advanced Composition Explorer, which sits in the solar wind ahead of the Earth, providing early warning of incoming solar storms.

Finally, orbital debris is the last space threat I will address today. The Joint Space Operations Center, managed by the U.S. Strategic Command, is tracking more than 23,000 objects in orbit around the Earth, of which about 95 percent represent some form of debris. In addition, millions of smaller debris objects that can potentially damage spacecraft are orbiting the Earth.

NASA continues to lead the world in studies to characterize the near-Earth space debris environment, to assess its potential hazards to the current and future space operations, and to identify and implement means to mitigate its growth.

In conclusion, NASA is making great progress in understanding and developing measures for mitigation of these space threats.

Again, thank you for the opportunity to testify today, and I look forward to responding to any questions you may have.

[The prepared statement of Dr. Green follows:]

PREPARED STATEMENT OF DR. JAMES GREEN, DIRECTOR, PLANETARY SCIENCE DIVISION, SCIENCE MISSION DIRECTORATE, NATIONAL AERONAUTICS AND SPACE ADMINISTRATION

Mr. Chairman and Members of the Subcommittee, I am pleased to have this opportunity to update the Committee on NASA's programs and our approach to addressing the risks, impacts, and solutions for space threats.

Orbital Debris

Today, the Joint Space Operations Center, managed by U.S. Strategic Command, is tracking more than 23,000 objects in orbit about the Earth, of which approximately 95 percent represent some form of debris. In addition, millions of smaller debris objects that could still potentially damage spacecraft are orbiting the Earth. For over 30 years, NASA has led the world in scientific studies to characterize the near-Earth space debris environment, to assess its potential hazards to current and future space operations, and to identify and to implement means of mitigating its growth. The NASA orbital debris program has taken the international lead in conducting measurements of the environment and in developing the technical consensus for adopting mitigation measures to protect users of the orbital environment. NASA is currently working to developing an improved understanding of the orbital debris environment and the measures that can be taken to control debris growth. NASA designs spacecraft to withstand the impacts of small debris and micrometeorites, and the Agency works with the Joint Space Operations Center to avoid collisions between our space assets and other known resident space objects.

Near-Earth Objects (NEOs)

NEOs are comets and asteroids that have been nudged by the gravitational attraction of nearby planets into orbits that allow them to enter the Earth's neighborhood. Composed mostly of water ice with embedded dust particles, comets originally formed in the cold outer planetary system while most of the rocky asteroids formed in the warmer inner solar system between the orbits of Mars and Jupiter. As the primitive, leftover building blocks of the solar system, comets and asteroids offer clues to the chemical mixture from which the planets and life eventually formed, making them compelling objects for scientific study.

The events of February 15, 2013, were a reminder of why NASA has for years devoted a great deal of attention to NEOs. The predicted close approach of a small asteroid, called 2012 DA14, and the unpredicted entry and explosion of a very small asteroid about 15 miles above Russia, have focused a great deal of public attention on the necessity of tracking asteroids and other NEOs and protecting our planet from them—something this Committee and NASA have been working on for over 15 years.

To put these two recent events in context, small objects enter the Earth's atmosphere all the time. About 100 tons of material in the form of dust grains and small meteoroids enter the Earth's atmosphere each day. Objects the size of a basketball arrive about once per day, and objects as large as a car arrive about once per week. Our Earth's atmosphere protects us from these small objects, so nearly all are destroyed before hitting the ground and generally pose no threat to life on Earth. While objects the size of the one that exploded over Russia (a rocky asteroid about 17 meters in diameter and weighing from 7,000 to 13,000 metric tons), enter the Earth's atmosphere roughly once every hundred years, they do remind us of the potential consequence of a larger impact. Even this small object resulted in about 6,000 buildings being damaged and about 1,500 people being injured, mainly from broken glass from the shock wave of the object exploding about 15 miles above the ground.

NASA leads the world in the detection and characterization of NEOs, and provides critical funding to support the ground-based observatories that are responsible for the discovery of about 98 percent of all known NEOs. NASA also has focused flight missions to study asteroids and comets. NASA uses radar techniques to better characterize the orbits, shapes, and sizes of observable NEOs, and funds research activities to better understand their composition and nature. NASA also funds the key reporting and dissemination infrastructure that allows for worldwide follow-up observations of NEOs as well as research related activities, including the dissemination of information about NEOs to the larger scientific and engineering community. Consistent with the President's National Space Policy, NASA continues to collaborate with the Department of Defense and other government agencies on planning and exercises for responding to future hazardous NEOs.

NASA was tasked by Congress in 1998 to catalog 90 percent of all the large NEOs (those of 1 kilometer or more in size) within 10 years; these would be large enough that should one strike Earth, it would result in a global catastrophe. NASA worked with a number of ground-based observatories and partners as part of our Spaceguard survey to reach that goal; NASA has now catalogued an estimated 95 percent of all NEOs over 1 km in size. None of these known large NEOs pose any threat of impact to the Earth anytime in the foreseeable future.

In 2005, Congress directed NASA to expand the survey to "detect, track, catalogue, and characterize the physical characteristics of near-Earth objects equal to or greater than 140 meters in diameter" (those that could destroy a city) and set a goal for this program to achieve 90 percent completion by 2020. NASA's NEO Observation Program (NEOO) currently funds three survey teams that operate five ground-based telescopes.

NASA has leveraged its investment in the Wide-field Infrared Survey Explorer (WISE) spacecraft by enhancing its operations to search for NEOs, resulting in the discovery of 146 previously unknown objects.

NASA's NEO Observation (NEOO) Program includes collaboration on ground-based telescopes such as the Space Surveillance Telescope with the Defense Advanced Research Projects Agency (DARPA) and the U.S. Air Force. NASA also funds the Panoramic Survey Telescope & Rapid Response System. The wide field of view survey capabilities of these two assets are expected to provide a significant increase in NEO detection rate.

However, ground-based telescopes will always be limited to the night sky and by weather. The only way to overcome these impediments is to use the vantage point of space. The privately funded B612 Foundation is planning to build a space observatory called Sentinel that would launch in 2018 and detect 100-meter sized objects and larger that could come near Earth's orbit. Sentinel will employ an infrared telescope from a Venus-orbit that will look "back" at the Earth in order to see and track near Earth objects. NASA is working collaboratively the B612 Foundation by providing technical assistance and operational support through a Space Act Agreement.

To find the more numerous smaller asteroids near Earth, NASA is initiating a project for development of an instrument that will be hosted on geosynchronous platforms such as communications, TV broadcast or weather satellites. This modest-sized, wide field telescope will have detectors that operate in the infrared bands where these faint asteroids are more easily detected against the cold background of space,

NASA is a leading participant in the NEO activities of the United Nations Committee on the Peaceful Uses of Outer Space (UNCOPUOS). Over the past several years, a working group on NEOs under the UNCOPUOS Scientific and Technical Subcommittee has been examining the topic of Earth-threatening NEOs. Results of that work led to recommendations this year, endorsed by the Subcommittee, to broaden and strengthen the international network to detect and characterize NEOs, and to call for relevant national space agencies to form a group focused on designing reference missions for a NEO deflection campaign. NASA is at the forefront of these activities and will continue to take on that role. NASA is also in discussions with our international partners to collaborate on several missions or mission concepts that could, in the future, grant additional access by U.S. researchers to valuable data on asteroids. NASA is working with the Japan Aerospace Exploration Agency or JAXA on potential collaboration on the Japanese-led Hayabusa II mission, building on our joint success with the earlier Hayabusa mission to the near-Earth asteroid Itokawa. NASA is also discussing with the European Space Agency potential collaboration on two of their mission concepts: (1) the Marco-Polo-R mission concept which is focused on returning a sample from a primitive near-Earth asteroid in the late 2020s, and, (2) the Asteroid Impact and Deflection Assessment (AIDA) mission concept that could be used to study the binary asteroid system Didymos with two spacecraft and see if a small interceptor can affect any the change in the relative orbit of the two bodies. Finally, the Canadian Space Agency launched their Near Earth Object Surveillance Satellite or NEOSSat in February 2013 to detect and track select near-Earth asteroids, and we look forward to seeing its data.

NASA is moving forward on the Agency's planned asteroid rendezvous and sample return mission, dubbed OSIRIS–REx (for Origins-Spectral Interpretation-Resource Identification-Security-Regolith Explorer), planned for launch in 2016. OSIRIS–REx will approach a near Earth asteroid, currently named 1999 RQ36 (one of the most exciting, accessible, volatile and organic-rich remnant currently known from the early Solar System), in 2019. After careful reconnaissance and study, the science team then will pick a location from where the spacecraft's arm will take a sample of between 60 and 1,000 grams (up to 2.2 lbs) for return to Earth in 2023.

Finally, NASA is working to accomplish the President's policy goal of sending an astronaut to visit to an asteroid by 2025. This mission, and the vital precursor activities that will be necessary to ensure its success, should result in additional insight into the nature and composition of NEOs and will increase our capability to approach and interact with asteroids.

Space Weather

Another threat from space being studied by NASA is space weather. Space weather refers to the conditions on the Sun and in the solar wind and near-Earth environment. Solar storms pose risks to humans in space and can cause disruption to satellite operations, communications, navigation, and electric power distribution grids; a severe geomagnetic storm has the potential to cause significant socioeconomic loss as well as impacts to national security.

Our ability to understand the Sun-Earth system is of growing importance to our Nation's economy and national security. The 2008 National Research Council report, *Severe Space Weather Events—Understanding Societal and Economic Impacts,* identified three industries whose operations would be adversely affected by severe space weather: electric power, space, and aviation. The electric power industry is susceptible to geomagnetically-induced currents, which can overload unprotected power grids and result in widespread power outages. With warning, power grid operators may be able to adjust operations to counteract such effects. In the spacecraft industry, intense geomagnetic and radiation storms have the capacity to disrupt normal operations such as satellite communication and television service. Space weather can cause irregularities in signals from Global Positioning System satellites. The aviation industry is susceptible to space weather events from both an operational and safety perspective. Communications between flights taking polar routes and air traffic control could be disrupted due to interference between the radio waves and the effects of space weather in the ionosphere. In addition, flight routes may be re-routed further south during solar weather events to reduce the radiation exposure to passengers and crew.

The National Oceanic and Atmospheric Administration (NOAA) is the official source for space weather predictions for the Nation. The U.S. Government, through the National Science Foundation and NASA, sponsors research programs to further our understanding of heliophysics and space weather. NASA's Heliophysics Division is responsible for formulating a national research program for understanding the Sun and its interactions with the Earth and solar system.

NASA currently operates 18 missions studying the sun and the solar wind, which have produced a number of scientific discoveries over the last year alone. Voyager has taken us to the edge of our solar system, the twin Solar TErrestrial RElations Observatory (STEREO) spacecraft have allowed us to view space weather events throughout the solar system, the Solar Dynamics Observatory (SDO) is helping us understand the causes of solar variability and its impacts on Earth, and the recently launched Van Allen Probes have already made new discoveries within Earth's radiation belts. Furthermore, for the first time, we have complete coverage of the Sun from all angles 24 hours a day, 7 days a week. We are now able to track the evolution of solar events from the solar interior to the surface of Earth, connecting the magnetized structure in the Sun's corona to the detailed features of Earth-directed coronal mass ejections (CMEs), or solar flares, to the intricate anatomy of geomagnetic storms as they impact Earth two to three days later. Several of these research satellites have become an essential part of our Nation's space weather prediction system. One example is the Advanced Composition Explorer (ACE) mission, which serves as an operational sentry for NOAA by providing early warning of incoming solar storms. However, ACE has been operating for 15 years and is well beyond its design life. Working with NOAA, NASA is refurbishing the Deep Space Climate Observatory (DSCOVR) in part to replace ACE's capabilities. Planned for a FY2015 launch, DSCOVR will have instruments that will provide critical operational space weather measurements to NOAA.

Conclusion

NASA's portfolio of missions and research addresses fundamental questions and at the same time, helps to protect our home planet from natural hazards from space. Research and early detection and evaluation of space threats are key to assessing the risks and providing critical information for mitigation to decision makers.

Again, thank you for the opportunity to testify today, and I look forward to responding to any questions you may have.

Senator NELSON. And you are going to put up, along with NOAA and the Air Force, Discover in late 2014 for giving us early warning on sun explosions, are you not?

Dr. GREEN. Yes, sir, that is correct. Indeed, Discover is a mission that is moving forward. It is a NOAA space weather mission, their first space weather endeavor at what we call L1, and it is indeed on track.

Senator NELSON. OK.

Dr. Lu?

STATEMENT OF DR. EDWARD T. LU, CHAIRMAN AND CHIEF EXECUTIVE OFFICER, B612 FOUNDATION

Dr. LU. Thank you, Mr. Chairman. I wanted to talk about asteroids and what we are doing about them.

The B612 Foundation is a Silicon Valley-based nonprofit that is building the Sentinel space telescope, and that telescope is going to find and track asteroids.

As we were reminded a couple of weeks ago, the Earth is sometimes hit by asteroids. The City of Chelyabinsk in Russia had an asteroid explode about—that asteroid was only about 18 meters across. That is about the—that would fit inside this room, roughly. Had an explosive energy about 25 times the bomb used in Hiroshima, or about 470 kilotons of TNT. The people of Chelyabinsk were very lucky. That asteroid exploded more than 40 miles from the city, and yet it still injured 1,000 people.

The last major impact before that was in 1908 in Tunguska. That asteroid was about 500 times the explosive energy of Hiroshima. It destroyed an area about the size of metropolitan D.C., and it was only about a factor of two larger. It would not quite fit in this room, but it is not much larger.

So I would like to start a video just to show you what the Solar System really looks like. This is an anatomically accurate depiction of all the known asteroids in our Solar System. So this is actually off of the JPL data base. This is every single known asteroid.

So the ones in the outside part of that, that is the asteroid belt. Those are the ones that won't hit the Earth. Those are the ones between Mars and Jupiter. The ones on the inside are the so-called near-Earth objects, and those are the ones that could hit Earth. And those are the ones that Dr. Green talked about. And those are the ones we are concerned with.

There are about 10,000 known near-Earth asteroids which have been discovered through NASA's Spaceguard program. However, we know that this is only a tiny fraction of those larger than the one that struck Tunguska that are out there. We know that there are about a million of them out there. We know this by counting craters and by knowing what small fraction of the sky we have actually been able to survey thus far from the ground.

So it turns out that, of these 10,000 that I am showing here— the light green line, by the way, is the orbit of the Earth. You can see how many of them fly past the Earth. There is actually about a factor of 100 more. For every one we know about, there are a 100 more we don't know about. And we simply don't know when the next one is going to hit the Earth because we don't know where they are.

The real situation looks more like this. This is what it looks like with a million near-Earth asteroids. And so you can see that the Earth really is flying around the Solar System in a cosmic shooting gallery, in some sense. And that is why these things hit the Earth.

So let me tell you a little bit about the odds of an asteroid hitting the Earth. You may be surprised by these odds. The odds of a 100-megaton impact this century—and 100 megatons, for scale, is about five times all the bombs used in World War II, including the atomic weapons. The chance of that in your lifetime, or this century, is about 1 percent on a random spot somewhere on Earth. Now, the odds of a much smaller 5-megaton impact, like we had in Tunguska, an asteroid that would not quite fit in this room, is about 30 percent. A 40 meter asteroid has a 30 percent chance.

So if I told you that there was a 30 percent chance of a random 5 megaton nuclear explosion somewhere on the surface of the Earth this century, what would we do to prevent that? And how is this situation any different?

Yes, most of the Earth is unpopulated, and we could get lucky. But wouldn't it be a shame if the area of the next impact was an area that wasn't unpopulated? And there are less and less areas that are unpopulated.

But there is good news. It turns out that we actually have the technology to deflect asteroids if we have a decade or more of notice. And NASA has been working on this; the National Academies has written a report on this. We understand how to deflect asteroids if we have advance warning. Because you can't deflect anything, you can't explore anything, you can't learn anything about something that you haven't yet found. That is the key.

So in the next video I will show you what the B612 Foundation is doing. We are building a space telescope; it is called Sentinel. And that is what it looks like. It is about the size of a FedEx moving van. And it is going to orbit the Sun, and it is going to track near-Earth asteroids.

According to the National Academies' findings, the best way to find asteroids is in the infrared, where asteroids are brightest, and from a vantage point where it can always look away from the Sun, so in an orbit around the Sun something like Venus.

So I will show you here where this thing is going to orbit the Sun, and that is where it can see—the white disc that you see is what it can see. The light-green line is the orbit of the Earth. So, as Sentinel moves around the Sun faster than the Earth does, it will scan Earth's orbit.

So it is going to find about 100 times more asteroids than all other telescopes combined. So it will be far, far and away more effective than all other telescopes combined. We have discovered, as Jim has said, about 10,000 near-Earth asteroids thus far with all of our telescopes over the last 30 years. Sentinel will discover roughly that number every 2 weeks. So it will be an impressive instrument.

We have assembled one of the world's finest spacecraft teams to work on this, and we have chosen Ball Aerospace in Boulder, Colorado, as our prime contractor. Sentinel is based on the design of the Kepler Space Telescope and the Spitzer Space Telescope, both of which were built by Ball. We launch in July 2018.

A couple things make this project unique. First, the B612 Sentinel project is being funded philanthropically. We are a nonprofit, and we will openly share the data with the world.

Second, we are managing this project in what I believe to be an innovative fashion, using commercial procurement practices. I proudly served at NASA for 12 years as an astronaut, and I also had the privilege of working at Google. And I think that we are using the best of both worlds in managing Sentinel. We are combining the technical rigor of NASA, which is the best in the world, with the innovative, rapid, and cost-effective practices that I learned at Google. And the secret to success, as I learned at both of these organizations, is hiring the very best people, and I think we have done that.

So I do want to talk a little bit about NASA. They have a very significant role in this. We are in a true sense a public-private partnership. We have a Space Act agreement in which NASA will be allowing us to use the deep space network of telescopes to transmit our data, and, also, NASA experts are part of our review teams. So they are a very important part of our project.

Sentinel will be important on a number of levels. So not only will it enable us to know if an asteroid is going to hit the Earth in time for us to actually deflect one, but it will find asteroids that merely come close to the Earth. And this happens all the time. And these asteroids that come close to the Earth will be attractive targets for exploration, both human and robotic, in the coming years.

So, should an asteroid be found on an impact trajectory with Earth—and I am reminding you that there is a 30 percent chance that there is a 5 megaton or larger impactor that is going to hit us this century—so should we find one, I believe that humanity will come together to prevent this. We will use our space technology to nudge this asteroid and prevent it from hitting the Earth. And I think that will be a watershed moment in human history.

And so thank you very much.

[The prepared statement of Dr. Lu follows:]

PREPARED STATEMENT OF DR. EDWARD T. LU, CHIEF EXECUTIVE OFFICER, B612 FOUNDATION

My name is Ed Lu, and I am the CEO of the B612 Foundation. Thank you for the opportunity to testify before the Senate Science and Space Subcommittee to describe the B612 Foundation Sentinel Space Telescope project. The B612 Foundation is a nonprofit 501(c) 3 organization dedicated to opening up the frontier of space exploration and protecting humanity from asteroid impacts. On June 28, 2012, the Foundation announced its plans to carry out the first privately funded, launched, and operated interplanetary mission—an infrared space telescope to be placed in orbit around the Sun to discover, map, and track threatening asteroids whose orbits approach Earth. Our name was inspired by the famed children's book by Antoine de Saint-Exupéry. B612 is the asteroid home of *The Little Prince*.

As the asteroid impact near Cheylabinsk Russia on February 15, 2013 vividly reminded us, our planet is occasionally struck by asteroids capable of causing significant damage. This was the largest asteroid impact since June 30, 1908, when an asteroid flattened 1000 square miles of forest in Tunguska, Siberia. The Earth orbits the Sun among a swarm of asteroids whose orbits cross Earth's orbit. These are *not* the asteroids that make up the asteroid belt between Mars and Jupiter, but rather the Near Earth Asteroids whose orbits take them much closer to the Sun, and who regularly cross the orbit of Earth. These asteroids are remnants of the formation of our solar system, and range in size from pebbles to many miles across.

More than a million of these Near Earth Asteroids are larger than the asteroid that struck Tunguska in 1908 with an energy more than 500 times greater than the atomic bomb dropped on Hiroshima. That asteroid was only about 40 meters across (about the size of a 3 story office building), yet destroyed an area roughly the size of metropolitan Washington, D.C. Unfortunately, less than 1 percent of the over one million asteroids greater than 40 meters have been identified to date. We therefore do not know when the next major asteroid impact will happen.

Currently there is no comprehensive dynamic map of our inner solar system showing the positions and trajectories of these asteroids that might threaten Earth. We citizens of Earth are essentially flying around the Solar System with our eyes closed. Asteroids have struck Earth before, and they will again—unless we do something about it. The probability of a 100 Megaton asteroid impact somewhere on Earth this century is about 1 percent. The odds of another Tunguska 5 Megaton event this century are much higher, about 30 percent. What if I told you there is a 30 percent chance of a random 5 megaton nuclear explosion somewhere on Earth this century? What would we do to prevent it?

But in the case of asteroids, we as a civilization have the capability to change the odds, and it is the mission of the B612 Foundation to ensure that such impacts do not happen again. Deflecting asteroids is technologically feasible, IF we have adequate early warning. If we know decades in advance of an impact, we can predict and actually prevent an impact using existing technology (kinetic impactors, gravity tractors, and if required, even standoff nuclear explosions) to nudge the asteroid and subtly change its course to miss Earth. Conversely, we can do nothing about an asteroid that we have not yet found and tracked. Thus, the first task we must undertake if we hope to protect ourselves from asteroid impacts is to conduct an astronomical survey of asteroids whose orbits approach Earth.

The B612 Foundation therefore decided to build, launch and operate a solar orbiting infrared space telescope called Sentinel to find and track asteroids which could impact Earth. Sentinel will be launched in July 2018, and during the first 6.5 years of operation will discover and track the orbits of over 90 percent of the population of Near Earth Objects (NEOs) larger than 140 meters, and the majority of those bigger than the asteroid that struck Tunguska (~40 meters). Sentinel will discover *100 times* more asteroids than have been found by all other telescopes combined.

Sentinel is novel amongst deep space missions in that it is being carried out by a private organization, the nonprofit B612 Foundation, and also because it is being managed using commercial practices under a milestone based, fixed price contract with the prime contractor Ball Aerospace and Technologies Corp. (BATC).

Sentinel Mission Overview

In 2005, the U.S. Congress recognized the need to extend the existing Spaceguard Survey for 1 km and larger NEOs down to smaller but still dangerous asteroids. The George E. Brown Act [1] authorized NASA to complete (>90 percent) a survey for NEOs down to a size of 140 meters, a size which while not threatening to human civilization is still capable of causing great damage (having an impact energy of roughly 100 Megatons of TNT). However, this future enhanced survey has not been funded by Congress, and the goal remains unfulfilled. Currently ~90 percent of NEOs larger than 1km have been discovered and tracked; while only about 5 percent larger than 140 meters, and only about 0.2 percent of those larger than 45 meters [2,3] have been tracked.

With this situation as a backdrop, the B612 Foundation decided in 2011 to undertake such a survey itself, and publicly announced the Sentinel Mission on June 28, 2012. Because asteroid deflection requires relatively small change in asteroid velocity when done many years to decades in advance of the impending impact,[4] the goal of this survey is to find and track asteroids with enough orbital accuracy to know if a serious threat exists and to give sufficient warning time to enable a successful deflection if necessary. We have chosen to adopt the 140 meter 90 percent completeness goal as our driving requirement, knowing that in addition in to generating a largely complete catalog at the 140 meter size level, many smaller yet still potentially dangerous asteroids will also be cataloged. The Sentinel mission is designed to give humanity sufficient warning time to be able to prevent threatening asteroid impacts.

Novel Private Funding and Commercial Program Management

One of the novel aspects of this mission is the way in which it is being funded. The B612 Foundation is a nonprofit charitable organization which is raising funds through philanthropic donations. Interestingly, large ground based telescopes (such as Lick, Palomar, Keck and Yerkes) have historically been largely funded through philanthropy.[5] In some sense Sentinel will be like these large observatories, with the exception that Sentinel will be in solar orbit rather than on a mountain-top. The B612 Foundation will in turn contract the spacecraft out to BATC, with B612 functioning in the role of program/contract manager and carrying out independent assessment of program progress. The total cost of the mission is currently under negotiation. The B612 Foundation expects to raise about $450M over the next 12 years to fund all aspects of this mission including development, integration and test, launch, operations, and program expenses.

The Sentinel mission is also taking an innovative approach to building and operating this interplanetary space mission. While previous missions that have departed from Earth orbit have been scientific investigations that have been developed with oversight by NASA or other governments (*e.g.,* ESA), Sentinel will be managed by B612 Foundation by adopting commercial practices for procurement and operations. Currently, communications and remote sensing imaging satellites (such as Digital Globe's WorldView series) are routinely procured under fixed-price contracts using commercial terms and conditions. These successful missions are compatible with such an approach because their performance requirements are very carefully specified in the contract and both parties are very familiar with the risks involved in the contract. In contrast, science missions typically push technology and performance margins in pursuit of innovative objectives. Furthermore, mission risks and possibly even the detailed design, are often not well understood at the time of contract signing. In these cases, NASA and contractors prefer a performance-based cost-reimbursable contract to limit the risk to the manufacturer. B612 has a very well-defined and stable requirement as articulated above. Thus, one of the prerequisites for commercial contracting is met.

One of the advantages B612 Foundation has as a private organization is that it is not bound by Federal procurement regulations. This allows B612 to make decisions and move quickly without the cumbersome regulations designed to prevent favoritism in Federal contracts, but which can add great overhead and slow decisions in cases where there is a clear best approach and contractor. BATC has carefully explored the implementation of the Sentinel mission and has identified high-heritage existing hardware system implementations (The Kepler and Spitzer spacecraft) that enable BATC to quantify the risk of manufacture and operation of Sentinel. Thus, we have been able to choose BATC as our contractor, and to make rapid progress towards a commercial contracting approach. This gives us the opportunity to enter into a fixed-price contract, an important feature for B612 since we must have a definite fund-raising target and do not have the ability to cover open-ended liabilities and cost-growth that might result from programmatic uncertainties. Crucially, the management of costs is the responsibility of BATC, which frees them from expensive accounting and compliance requirements associated with cost-reimbursable contracts.

A key feature of a successful implementation of this commercial contracting approach is frequent and detailed communications between B612 and BATC. While BATC is responsible for meeting performance requirements, B612 remains aware of programmatic risks and mitigations and approves the progress of the work. This is facilitated by the identification of milestones within the contract that detail various development achievements at which point the progress of the overall contract can be assessed. These assessments provide opportunities for dialog on programmatic and mission risks and mitigations. This arrangement is relatively hands off compared to typical large space missions, and works both because B612 has a small but highly experienced technical team, and because of the high heritage of the BATC design. B612 has also enlisted an independent panel of experts known as the Sentinel Special Review Team[6] to provide advice on technical and programmatic risk to B612. In addition, B612 will have permanent on-site technical and management personnel to enhance our visibility into progress on the contract. This approach has been implemented with great success on numerous other commercial space missions.

Another key aspect of the mission is support from NASA. B612 Foundation and NASA have signed a Space Act Agreement[7] in which NASA will provide use of the Deep Space Network (DSN) for telemetry and tracking, as well as allowing NASA personnel to participate on the independent technical advisory team known as the Sentinel Special Review Team. NASA and the scientific community benefit because B612 will make the data available to the community through the standard process for reporting NEO observations (see Detection Scheme below).

Sentinel Mission Overview

The Sentinel mission places an infrared imaging telescope in a Venus-like orbit to identify and catalog NEOs over a 6.5-year mission life. Figure 1 shows Sentinel's viewing geometry. The "Venus-like" orbit at ~0.7 AU provides up to a 200 degree, anti-sun viewing field that the observatory methodically scans to detect the infrared light coming from any moving object in the field. By making observations from ~0.7 AU, Sentinel views a much larger portion of the sky relevant to finding NEOs than can be seen from the Earth, either from ground-based or space-based observatories. A space-based survey is also not compromised by the atmosphere, or by the presence of the moon, or by the requirement to look for NEOs low in the sky during twilight. Locating Sentinel in space near 0.7 AU from the Sun has the additional benefit of being interior to most NEOs, thereby observing them when they are closest to the Sun and at their brightest. This, or a similar orbit, is essential for detecting those long-synodic-period (low relative velocity with respect to Earth) NEOs that are the most dangerous and valuable to future exploration missions.

Sentinel will be launched from Earth on a Falcon 9 rocket. The cruise to the final heliocentric orbit at ~0.7AU uses a Venus gravity assist to minimize fuel requirements. Communications through the DSN with Sentinel consists of two kinds of interactions. Infrequent command uplinks occur through low speed command link, while mission data uses a high speed downlink. The total downlink data volume is ~4 gigabits a week, and the DSN link-time is approximately 4 hours per week. Flight data from the DSN is first processed at a ground station at the Laboratory for Atmospheric and Space Physics at the University of Colorado.

The Sentinel uses proven designs successfully flown on the Kepler and Spitzer missions to demonstrate feasibility and low development risk, and to provide a firm cost basis. Figure 2 shows the notional Sentinel Observatory. The tall structure on left is the thermal shield, which also carries the body-fixed solar array, a system based on *Spitzer*.[8]

The central region shows the two intermediate-temperature thermal shields, rendered in brown. To the right of the intermediate temperature shields is the 50-cm-aperture mid-wave infrared (MWIR) telescope. The telescope is cooled to 45K by a combination of radiative and active cooling. The instrument's HgCdTe focal plane is actively cooled to 40K. The detection band from 5 to 10.4 microns is optimized for detecting T=250K objects, a characteristic temperature for NEOs near 1AU. The telescope is mounted on a *Kepler*-derived spacecraft, and reuses *Kepler's* avionics and structure.

Detection Scheme

To detect a NEO, we require two pairs of observations of the anti-Sun hemisphere in 24 days. The basic detection scheme's timeline is presented in Figure 3. There are 4 separate observations made of every part of the anti-Sun hemisphere every 24 days (and 4 pairs on most sections in 26 days.) The images are taken in correlated pairs that reveal the motion of any NEO in the 1-hour span between images. All the data for each pair of images is first stored, and then later compared onboard, and NEOs are detected by their motion during the one hour interval between the two images.

We greatly reduce the amount of telemetry data by retaining only those portions of the imaged field that contain pixels determined by the dedicated on-board Payload Computer to contain moving objects. Additionally, for each tile we include roughly 100 well-known infrared stars used to establish an astrometric grid.

At the ground station, these observations are converted into detection-fragments called "tracklets." Tracklets are then sent to the IAU Minor Planet Center (MPC) in Cambridge, Massachusetts. The MPC maintains the world's NEO database, and will convert tracklets into orbits. These MPC orbits then go to the Near-Earth Object Program Office at JPL which refines the initial MPC orbits, calculates the likelihoods of any impacts and globally distributes its findings.

Survey Performance

In 6.5 years of operation, Sentinel will detect and track the great majority (>90 percent) of all NEOs larger than 140m. In addition, Sentinel will detect and track 50 percent of all NEOs greater than 50m. Figure 4 presents Sentinel's NEO cataloguing rate. These results were generated using an integrated-systems model which includes a modeled NEO population in combination with spacecraft telescope and detector performance models as well as the preliminary observing cadence described in figure 3. We iteratively used the model to guide our design through the phase-space of options until we hit the 90 percent level for 140 meter objects on this plot. Among the parameters considered in these trade studies were aperture, field of view, detector wavelength cutoffs, final spacecraft orbital parameters, focal plane array operating temperature, detection thresholds, pixel size, integration time, etc. Over 75 such trade studies have been carried out thus far.

Summary

Sentinel is important on a number of levels. First, the B612 Foundation is pioneering a new model for carrying out large space missions in which Sentinel is philanthropically financed and privately managed, but with a crucial government partnership. Second, the primary goal of the mission is not scientific. While it is true that Sentinel will be a groundbreaking new astronomical instrument, the primary requirement for the mission stems from a planetary defense (*i.e.*, public safety) goal. Once Sentinel is in operation, it will generate a flood of new NEO discoveries, far in excess of all other observatories combined. After 6.5 years of operation it will discover and track approximately 1,000,000 NEOs, as compared to the currently known total of about 10 thousand. Not only will this catalog provide a list of potential targets for robotic and human exploration, but should any of these NEOs be on a collision course this information can allow us to successfully mount a deflection campaign and prevent a catastrophe. Our future may depend on it.

References

1. *http://www.gpo.gov/fdsys/pkg/PLAW-109publ155/pdf/PLAW-109publ155.pdf*
2. Harris, A.W. NATURE 453, 1178–1179, 2008.
3. Alan W. Harris, personal communication
4. Schweickart, R.L, 2009. *http://dx.doi.org/10.1016/j.actaastro.2009.03.069*
5. MacDonald, A. 2012. PhD Thesis.

6. *http://b612foundation.org/the-foundation/sentinel-review-team/*
7. *http://b612foundation.org/images/SAA__redacted.pdf*
8. M. W. Werner *et al.* 2004 *ApJS 154* 1 *doi:10.1086/422992*

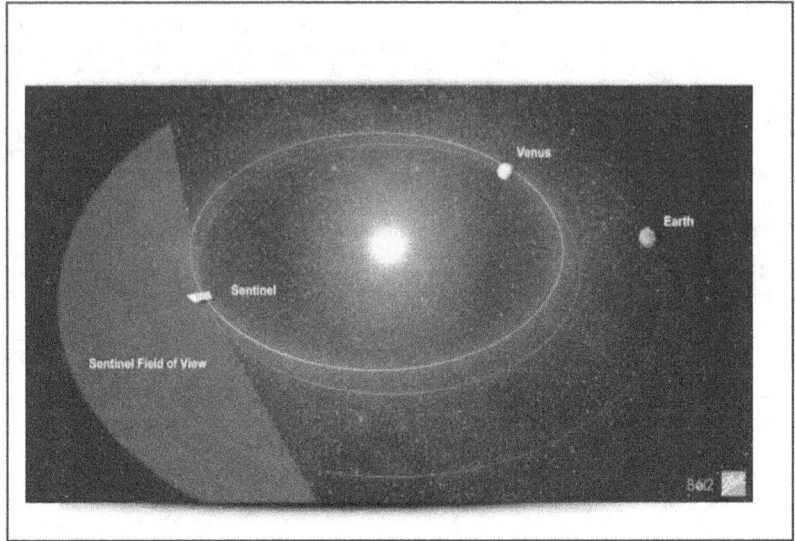

Figure 1—Sentinel's mission architecture enables it to detect and track NEOs within a much larger search volume than is available from the ground, and without the constraints of weather and lunar cycles.

Figure 2 shows the Sentinel Observatory. It consists of a rebuild of the *Kepler* spacecraft (modified for the Venus-like orbit) and an infrared telescope.

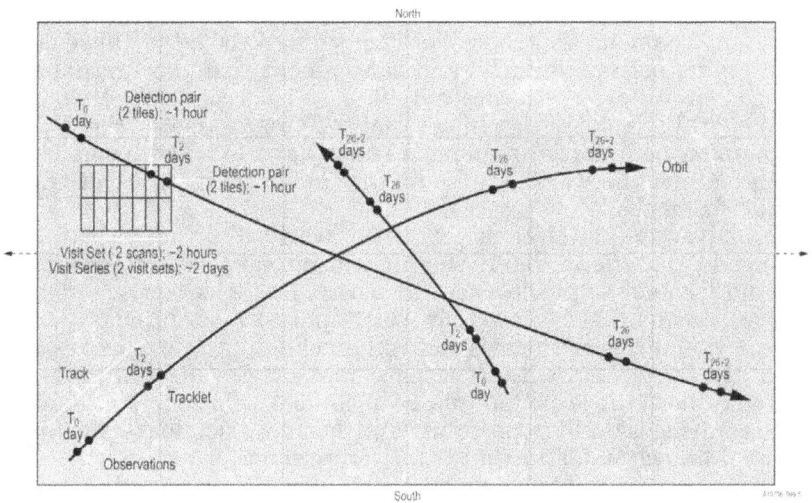

Figure 3. The basic viewing scheme uses one-hour pairs on two-day and then 26-day centers to locate moving NEOs.

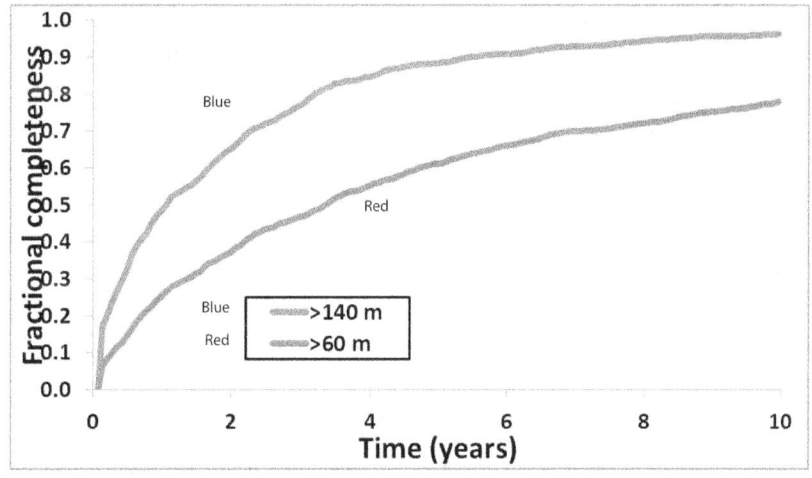

Figure 4. Survey completeness for all NEOs > 140-meter and 60-meter diameter vs. time for Sentinel, assuming simultaneous operation of ground based Pan-STARRS1

Senator NELSON. And we want to get into how you are going to nudge it away. And we will get into that.

Mr. DalBello?

STATEMENT OF RICHARD DALBELLO, VICE PRESIDENT, GOVERNMENT AFFAIRS, INTELSAT

Mr. DALBELLO. Thank you, Mr. Chairman, Senator Cruz. It is a pleasure to be here today to talk about the issues of space risks and how they influence the commercial operators who are earning their living day to day in space.

Intelsat has been in this business for about 50 years. We are currently flying over 70 satellites. So we are pretty familiar with the space environment and the risks it entails. As a global fleet operator serving both commercial and government customers, reliability and continuity of service are our highest priorities. Whether it is UAV operations over Afghanistan or the final game of the NCAA tournament or financial statements that have to be transferred securely around the world, we know that our customers expect flawless performance.

To deliver this level of performance, we have to daily deal with a range of threats. Probably the highest priority issue for us today is radio frequency interference. In most cases, this service disruption is accidental, however, it sometimes is intentional. Other threats include space debris and other challenges of space flight; cyber attacks; solar weather; space systems reliability; the fact that we today don't have an affordable technical solution for refueling and repairing satellites on orbit; and last but not least, an international launch industry that is far from robust.

Now, our economy depends on the ability to create and instantly distribute vast amounts of data around the planet. Space-based platforms have become a vital link in the national and global economies. They are essential to the prediction of weather, navigation in all kinds of transportation, the operation of power grids, the completion of local and global financial transactions, and communication to mobile platforms, whether they be on land, sea, or air.

The commercial satellite industry also plays a critical role in supporting government operations. Commercial satellites supply the majority of communications in Afghanistan and Iraq. Today, our satellites are still flying almost all of the DOD's unmanned aerial vehicles, and we are providing the vast majority of the Navy's communications at sea.

To address the challenges that I mentioned earlier, the leading space operators have gotten together on a number of complex cooperative projects.

Probably the most significant of these is the Space Data Association, or SDA of which Intelsat is a founding member. The formation of SDA is a major step toward creating a voluntary space traffic control, if you will. It is an interactive repository of satellite orbit and maneuver information and soon will contain satellite configuration data. That will allow us to also use the same database to help resolve radio frequency interference issues.

The Space Data Center allows satellite operators to augment government-supplied data with precise orbit data, maneuver plans, and to retrieve information from other member operators when necessary.

Determining the orbit of objects in geo is a complicated task. The U.S.'s current Space Surveillance Network is subject to a number of constraints. Some of those constraints are weather, scheduling, geographic diversity, and overall capacity. Recent efforts within SDA to share owner-operator data provide clear proof of the value of collaboration. NASA and NOAA have both joined SDA and are providing information into this common data base.

In creating SDA, the private sector has taken the first step toward a new paradigm of managing risk in space, but to be most effective, far more cooperation is needed.

In other areas of interest to this committee and being discussed today, such as space weather or highly uncommon events, such as the asteroid fly-by, there are no clearly established links between the government and the commercial sector.

In space weather, although we are aware of the good work that NOAA is doing at the Space Weather Prediction Center, there is no established alert protocol between government and industry. Nor is it clear what levels of solar activity would mandate change to routine operations. So it is one thing that good solar information exists, but the lack of a communication ability to translate that information into action is a significant deficit.

For a highly uncommon event, such as an asteroid fly-by, there is simply no established communication mechanism. I believe our flight operations team learned of DA14 when they received a courtesy call from a colleague at the Aerospace Corporation.

Last year, the commercial satellite industry participated in DOD's Schriever Wargames. Now, these games are held every other year, and they are designed to exercise DOD thinking about the deployment of its terrestrial and space assets in response to a conflict situation. Last year, those games concluded, as they have several times in the past, that DOD relies on the commercial satellite industry—their reliance is considerable and that a crisis is the wrong time to try to establish clear lines of communication with your major partners and suppliers. I suspect the same conclusion can be safely applied to the topics that we are discussing today.

While governments were first to send satellites to near-Earth space, commercial enterprise will be the primary user of the orbital arc in the 21st century. Governments and commercial space operators need to take a more collaborative approach to enhancing the safety of the space environment. The SDA is an important step on this path.

With the support of the U.S. Government, we can create an international framework that acknowledges the vital contribution of commercial industry while working to assure the preservation of the space environment.

Thank you, sir.

[The prepared statement of Mr. DalBello follows:]

PREPARED STATEMENT OF RICHARD DALBELLO, VICE PRESIDENT, GOVERNMENT AFFAIRS, INTELSAT

Good morning. I am Richard DalBello, Vice President of Government Affairs for Intelsat.

It is a pleasure to be here today to discuss the "Risks, Impacts, and Solutions for Space Threats" as they pertain to the commercial satellite industry. As requested, I will also comment on the state of the current collaboration between industry and government on these topics and offer some ideas on how to improve the planning and cooperation between the U.S. Government and the industry.

Intelsat is the world's leading provider of satellite services. With almost 50 years of service and over 50 satellites in orbit, we are familiar with the complex challenges of this industry. As a global fleet operator, our concerns go beyond the operation of individual satellites but are, instead, focused on the maintenance of a highly sophisticated global fleet of satellites and our IntelsatOne terrestrial network, providing a multiplicity of services to large media, corporate, and government customers.

Continuity of service is one of the highest priorities of our commercial and government customers. Large media companies, broadcasters, global corporate networks, and government users must feel that our satellite services are dependable and that their critical services will not be interrupted. Thankfully, today's satellites are highly reliable and they often outlive their notional 15-year lifetimes. Of course, anomalies can occur and satellites must be replaced at the end of their useful lives. Maintaining a fleet of over 50 satellites means that we are launching several replacement satellites each year. This raises two other significant topics: the importance of a vibrant, domestic launch industry and the relevance of investments in next-generation technologies to allow the refueling and repair of satellites on orbit. Although these topics are beyond the scope of this hearing, they have a significant impact on our country's current and future ability to respond to space threats.

Increased Reliance on Satellite Communications

Over the last several decades, the U.S. economy and the Federal Government have both grown increasingly reliant on the commercial satellite communications industry. Today, such vital activities as television broadcasts, the Internet, oil and gas exploration and production, financial transactions and agricultural production all depend, in part, on the ability to communicate by satellite.

Our economy now depends on the ability to create and instantly distribute vast amounts of information around the world. Space-based communications platforms have become vital to the day-to-day linking of national and global economics, the prediction of weather, the navigation of virtually all forms of transportation, the operation of power grids and the completion of local and global financial transactions. In remote parts of the globe, satellites provide the only link to more populated areas. A sudden loss of satellite communications would cause significant economic disruption.

The commercial satellite industry also plays a critical role in supporting government operations, including national security and emergency preparedness missions. The commercial industry supplied the majority of the satellite communications used for military operations in Afghanistan and Iraq and continues today to provide nearly all beyond-line-of-sight communications for our unmanned aerial vehicle (UAV) fleets.

Commercial Satellite Vulnerabilities and the 2009 NSTAC Report

In 2008, Intelsat participated in a review by the President's National Security Telecommunications Advisory Committee (NSTAC) to identify both physical and cyber security threats facing the commercial satellite industry, mitigation measures employed to combat such threats, and initiatives to develop a standard security framework among satellite operators to enhance national security.[1] The report was published in 2009, but its major conclusions are still relevant today.

Among the NSTAC Report's conclusions were:

- *Radio Frequency Interference (RFI)*—Radio frequency interference represents a significant and growing threat to satellite services, yet Government and industry do not collaborate systematically to share information regarding the detection, characterization, geolocation,[2] and mitigation of interference. The Government engages with industry only when a Government service is affected instead of working collaboratively with industry to identify best practices and establish shared situational awareness and mitigation approaches.

- *Cyber security*—The terrestrial components of satellite networks contain many of the same subsystems found in other communications networks. As a result, satellite and terrestrial networks share similar cyber vulnerabilities and mitigation measures. However, because satellites must be controlled remotely from Earth, satellite operators take special care to mitigate two risks: (1) remote introduction of a false spacecraft command; and (2) a malicious third party preventing the spacecraft from executing authorized commands or interfering with satellite telemetry reception. Consistent with Government policy, most satellite companies use the National Security Agency-approved satellite command uplink encryption for satellites supporting U.S. Government services.

---

[1] NSTAC. (2009). *NSTAC Report to the President on Commercial Satellite Communications Mission Assurance.*

[2] Geolocation is a technique that allows satellite operators to rapidly identify the location of an interfering signal by using advanced signal processing techniques coupled with other known information.

- *Space Traffic Control*—While an accidental collision between space debris and a satellite is unlikely, collisions do occur, can be catastrophic, and can cause permanent damage. The February 2009 collision of an Iridium communications satellite and a defunct Russian Cosmos satellite provides one example. Every such collision produces additional debris that remains in the space environment, often for years, and poses an ongoing threat to other spacecraft. Preventing collisions is of paramount importance. The NSTAC found that, today, the Department of Defense (DOD) shares only limited space situational awareness information with private industry. However, promising initiatives such as industry's Space Data Association should promote better location sharing, maneuver coordination, and collision avoidance.
- *Protection of Terrestrial Infrastructure*—Satellites are far less likely than terrestrial facilities to be the target of a successful physical attack due to their location in space. The NSTAC found that satellite operators use redundant and geographically diverse facilities to protect terrestrial infrastructure from man-made and natural threats to ensure continuity of critical satellite network functions. Ground stations are connected by redundant, path-diverse, cryptographically secured communications links and employ preventative measures such as buffer zones and robust security systems to protect from attack. Further, operators maintain personnel security procedures, including background checks, employee badges, logged entry and exit, and on-site security guards, as part of their best-practice security efforts.
- *Collaborative Forums for Government/Industry Dialogue*—The NSTAC report noted, with approval, the creation by DOD of the Mission Assurance Working Group (MAWG) to encourage a constructive and collaborative relationship between DOD and the satellite industry, including at the classified level. The MAWG had undertaken a variety of issues including enhancing compliance of commercial services with DOD mission assurance requirements, increasing mission assurance through modifications and improvements to communication architectures, and suggesting new or revised capabilities for commercial service acquisitions.[3]
- *Long Term Planning*—Satellite operators make every effort to replace existing satellites with updated or enhanced systems to meet both future commercial and Government user requirements. However, the Government does not engage with industry in planning for its long-term communications needs. As a result, the Government relies on the "spot market" to meet most long-term service needs and risks a potential shortfall in commercial satellite availability when critical needs arise.
- *Space Weapons*—Due to the technological availability and/or cost of mitigation, the commercial satellite industry does not mitigate the risk of nuclear detonations or space weapons.

Space Data Association—Industry Collaboration on Safety of Flight[4]

Since the launch of Sputnik in 1957, governments and commercial companies have placed thousands of satellites in orbit around the Earth. Most of them have long since burned up reentering the atmosphere or disintegrated into space debris. Today, there are still more than 16,000 active satellites and debris objects in the public catalog of tracked objects.

The region of space near Earth in which satellites orbit is so large—extending out 22,200 miles for commercial satellites—that one might believe a collision of orbiting spacecraft would be impossible. However, just four years ago, a satellite operated by Iridium Communications for the company's global communication network collided with an uncontrolled Russian spacecraft that had been out of service since 1995. The collision, 490 miles above Siberia, produced over 2,000 pieces of debris larger than 10 centimeters (3.9 inches) in diameter, each one large enough to destroy any orbiting satellite in its path.[5]

To avoid collisions in the increasingly crowded orbital arcs, agencies and companies operating satellites have informally shared position and orbit data for many years. One problem with this informal information sharing is that satellite operators

[3] Since the publication of the NSTAC report, the MAWG has been disbanded. Discussions are underway between DOD and industry to replicate some of the functions of the MAWG but, to date, no formal structure has been established.

[4] Some of the material in this section was previously published. See: DalBello, Richard. (2011). Managing Risks In Space. Federation of American Scientists. Retrieved from *https://www.fas.org/pubs/pir/2011winter/2011Winter-ManagngRiskinSpace.pdf*.

[5] NASA Orbital Debris Quarterly News, July 2011.

don't use the same standard to represent the position of a satellite in orbit or an object in space. Many different types of software are used to track and maneuver satellites and the data is stored in a variety of formats. So even operators who wish to share data can't rely on a single, agreed-upon protocol for sharing information. As a result, operators sharing information must maintain redundant file transfer protocols and tools to convert and reformat data so that it is consistent with their own software systems to compute close approaches. As the number of satellite operators increases, the problem of maintaining space situational awareness grows more complex. And the smallest operators may not be able to afford, or have the technicians, to participate in the data sharing process.

Recently, the world's leading commercial satellite operators formed the Space Data Association (SDA) to formalize the process of exchanging information and to deal with the overall data compatibility problem. One way to minimize risk in space is for all operators to share what they know about the movement and position of their own satellites in a way that all other companies can use. While this sounds like common sense, governments and commercial companies around the world have each historically acted predominantly on their own in launching and monitoring satellites. Agencies and companies coordinate frequency allocation and orbital slots prior to launch, but once a satellite is in orbit, data about the movement of commercial satellites was shared only informally until the establishment of the SDA. Information about the operation and location of many military and intelligence satellites is still shrouded in secrecy.

The most critical times to share data about satellites are when a new satellite is being placed in orbit or an existing satellite is being shifted from one orbital slot to another. A typical communications satellite is as big and heavy as a loaded semitrailer, and though it appears fixed above the Earth, it is actually traveling thousands of kilometers per hour. Putting a satellite into an orbital slot or moving it to another position above Earth without disturbing any of the other 250+ commercial communications satellites in the GEO[6] plane, as Intelsat routinely does, is a very delicate operation. Yet this process is managed entirely by commercial operators using informal, de facto rules developed through experience and implemented by consensus.

The formation of the SDA is a major step toward creating a voluntary "space traffic control" system for space. The SDA is an interactive repository for satellite orbit, maneuver, and payload frequency information.[7] The SDA's principal goal is to promote safe space operations by encouraging coordination and communication among its operator participants. Through the SDA's Space Data Center, the satellite operators maintain the most accurate information available on their fleets; augment existing government-supplied data with precise orbit data and maneuver plans; and retrieve information from other member operators when necessary. As a result, the data center:

- *Enhances Safety of Flight.* The SDA aims to preserve the space environment by rapidly and automatically sharing information about the positions of satellites in space.
- *Reduces Radio Frequency Interference.* Radio frequency interference—both intentional and accidental—is the number one operational problem facing communication satellite companies today. By sharing the precise location of commercial satellites and the configuration of their payloads, operators can more rapidly find and address interference sources.
- *Simplifies Communication in a Crisis.* Before creation of the SDA, the world's satellite operators had no authoritative index of contact information for engineers actually controlling another company's satellites. Although there was always a great deal of informal communication, the SDA has standardized and automated the information necessary to communicate between technicians in operations centers during a crisis.

Because of the proprietary nature of the operational data, the SDA has been designed to protect information and prevent participants from using for commercial purposes the data supplied by other operators. The participants of the SDA contribute operational data through a secure interface on a daily basis and can access data related only to the operation of their own satellites. For example, an operator

---

[6] Most commercial and military satellites operate in one of two orbit planes. The first, low-Earth orbit (LEO), is between 160 and 2,000 meters (100–1,240 miles) above Earth's surface. The other, geostationary Earth orbit (GEO), is a circular orbit 35,786 kilometers (22,236 miles) above the equator.

[7] See: *www.space-data.org.*

who only has satellites covering Latin America cannot access data from other parts of the globe.

So far, the SDA has 21 contributing operators and maintains precise position information on 267 satellites in GEO, and another 90 satellites in LEO. Additionally, both NASA and NOAA joined the SDA in 2012. The greater the participation of the SDA, the more comprehensive the data and the resulting analysis will be. As new satellite operators continue to join the SDA, the data center will continually improve its reliability in all satellite arcs and develop into a truly global and comprehensive database for space situational awareness.

Several years ago, the U.S. Government began providing the public, including satellite operators, with satellite position data gathered, using radars and sensors, by the U.S. Strategic Command (USSTRATCOM). The position information provided initially for close-approach monitoring, called two-line element (TLE) data, had several drawbacks. First, there was no fixed standard for TLE interpretation. Second, TLE data did not have the required accuracy for credible collision detection. Recently, USSTRATCOM developed a procedure for providing satellite operators with more comprehensive information in the form of conjunction summary messages (CSMs). These CSMs are used to warn operators whose satellites have been identified by STRATCOM as closely approaching another space object.[8] These CSMs contain vector and covariance information computed from other data, making them more accurate than TLEs.

However, recent studies funded by Intelsat and SES have concluded that to ensure the highest level of accuracy, it would be beneficial for USSTRATCOM to incorporate data from routine satellite maneuvers. The SDA has offered to augment the global data maintained by USSTRATCOM with more precise operator-generated data to improve the accuracy of conjunction monitoring. The SDA could also provide a standardized method and focal point for operators to share information and facilitate communications between satellite operators and governments interested in making available timely space object catalogues. Hopefully, with the passage of time, the U.S. and other governments will be able to fully capitalize on this industry-sponsored and funded initiative. Solving the problem of government/industry data sharing and the role of the SDA should be a key objective of future international discussions on this topic.

Another major risk to operators is the proliferation of orbital debris from rocket stages, defunct satellites, equipment lost by astronauts and the fragments left from explosions and collisions of satellites. For example, Vanguard 1, launched by the United States in 1958, is expected to remain in orbit at least another 200 years before slowly burning up as it drifts down into the atmosphere.[9] The debris problem is most severe in low-earth orbit (LEO), where the majority of satellites used for communications and remote sensing operate. Because these satellites are not geostationary, multiple satellites, rather than a single satellite, are required to provide continuous coverage of any given area.

While governments were the first to send satellites to near-Earth space, commercial enterprises and consumer services will be the primary users of the orbital arcs in the 21st century and, hopefully, beyond. Consequently, governments and companies operating spacecraft need to take a more collaborative approach to enhancing the safety and efficacy of the space environment. The Space Data Association is the major step on this path, and that step should be followed by firm actions of governments and all space users to create an international framework that assures the preservation of this valuable resource.

Radio Frequency Interference

Radio frequency interference (RFI) is a serious problem that costs the satellite industry millions of dollars each year. The users of satellite services routinely state that RFI is the single most important issue relative to their use of satellite services.[10] RFI disrupts television signals, data transmissions and other customer services, requiring significant operator resources and hindering business growth. Interference has a financial impact as well to satellite operators and users. When there is interference on a satellite, there is revenue lost due to the reduction of available bandwidth and power capacity. Expenses are increased, ranging from the purchase

---

[8] Statement of Major Duane Bird, USAF, U.S. Strategic Command to *AMOS Conference,* September 2010.

[9] NASA's National Space Science Data Center.

[10] Intelsat. (2012). *Carrier ID Wins a Gold Medal at the 2012 Summer Olympics.* [Blog]. Retrieved from *http://www.intelsatgeneral.com/blog/carrier-id-wins-gold-medal-2012-summer-olympics.*

of interference monitoring or geolocation equipment to hiring and dedicating personnel to interference mitigation.

Intelsat has played a lead role in global efforts of commercial satellite operators to foster an interference-free space environment. Intelsat is working with satellite operators, industry groups, customers and equipment manufacturers to make RFI reduction a top priority.

There are both long-and short-term causes of interference.[11] Long-term interference typically occurs between two adjacent satellites and can be caused by lack of coordination between users, outdated or poorly designed equipment, or small mobile antennas. Terrestrial sources, such as microwave links or radar signals may also cause long-term satellite interference. Although it has been rare, interference can also be the result of deliberate, politically motivated actions, such as the recently-reported Iranian jamming of certain western broadcasts. Short-term interference typically results from poor training and operator error. Over 80 percent of interference events experienced each year result from some form of user error. Proper training is critical for reducing RFI incidents. A majority of RFI incidents are attributed to faulty installation practices, uplink errors and poor equipment maintenance regimes. Intelsat and other leading operators are endorsing new, comprehensive training and certification programs to educate technicians on proper equipment installation and operational parameters.

One concept recently embraced by the global satellite industry is the deployment of "Carrier ID" technology to help identify the interference source. Carrier ID is a stamp on uplink signals that enables satellite operators to more efficiently trace the source of transmissions to their satellites and thereby speed the remediation of any signal interference. Carrier ID would be on every carrier transmitted to the satellite. It is a small identification that may include the operator name, the contact's telephone number, or the modem serial number. The goal is that, at any given monitoring location, a single system can extract the Carrier ID for any and all carrier types where Carrier ID insertion has been provided. This will allow satellite operators to communicate directly with the RFI source to resolve the incident.

The 2012 Summer Olympics in London were the most-watched television event in U.S. history, attracting over 219 million viewers over 17 days of coverage. The London Games were also a perfect opportunity to test Carrier ID technology. The major satellite providers all deployed Carrier ID with positive results.

In addition, the Olympic experiment served as a test bed for a Carrier ID Database. This database was developed in partnership with SDA as an adjunct to the existing work of the organization. The open exchange of operational data is imperative for critical satellite operator procedures, including RFI identification, analysis and RFI geolocation. The Carrier ID Database was designed to be complementary to the other services of the SDA. When implemented, the SDA could then provide a central repository where satellite operators can standardize, formalize and automate data collection.

Although the commercial satellite operators have devoted a considerable amount of time and resources to the issue of RFI and although this issue is of high importance to the U.S. Government, there has been very little real coordination on this topic or the larger spectrum topics that face all satellite users. Both DOD and industry are under pressure in the U.S. and around the world to release valuable spectrum, or to share spectrum, with the fast-growing terrestrial wireless industry. As stated above, DOD relies on the commercial sector for the vast majority of its satellite communications requirements, including virtually all of its beyond-line-of-sight UAV communications. The commercial satellite operators have made a number of proposals for more creative sharing regimes, such as hosted payloads and DOD spectrum-specific commercially operated satellites. To date, DOD has been reluctant to embrace any of these forward-leaning proposals.

Solar Weather Effects on Satellites

The sun is the dominant element in the determination of "space weather" and satellite operators monitor the sun's activities in order to improve their ability to respond to the impact of solar events.[12] There is, occasionally, speculation that the partial or complete disabling of a communications satellite might have been the result of a solar effect. Such reports are the result of analysis and speculation since physical analysis of the satellite is impossible. For the most part, satellite manufacturers build their products to operate in the sometimes-harsh environment of space

---

[11] *Carrier ID Using MetaCarrier®* Technology, ComTech white paper, *http://www.comtechef data.com/*

[12] Intelsat. (2013). Solar Weather [White Paper]. Retrieved from *www.intelsat.com/tools-resources/satellite-basics/solar-weather/*

and satellite operators have come to assume that their satellites will withstand such "weather" events. The goal of satellite fleet operators has been to identify and effectively counter the sun's link to so-called single-event upsets (SEUs), which happen whenever the performance of one or more spacecraft components abruptly changes without warning.

Solar researchers, space weather forecasters and satellite operators focus on four elements of solar weather that can affect satellite communications: solar wind, coronal holes, coronal mass ejections (CMEs) and solar flares. The solar wind is constant but varies in intensity, while the other three solar phenomena are more highly variable. SEUs are not apt to be caused by the solar wind itself, which is relatively low in energy and seldom penetrates the outer layers or protective skin of a spacecraft. Instead, coronal holes, CMEs and solar flares can be more potentially disruptive. When solar storms erupt, they can bombard a satellite with highly-charged particles and increase the amount of charging on the spacecraft's surfaces.

Coping with electrostatic discharges from the sun that can potentially disrupt satellite services are part of the everyday reality of the satellite world. Losing solar power is not a serious concern whereas losing total control and command of a satellite as the result of solar weather is the most severe effect. Solar panels on satellites are the most affected components, and normal erosion rates for solar panels are usually 0.3 percent to 1 percent per year. A solar storm can reduce solar panel performance by 3 percent to 5 percent in a day, but since this phenomenon is well understood, spacecraft manufacturers increase the tolerances by design, and attach larger than needed solar panels to a satellite in order to allow for losses during the anticipated solar storms.

The body of a communications satellite, which contains vital control and communication components, is built with special materials as well as active and passive measures so as to be highly resilient to the sun's effect. A so-called "Faraday Cage" protects the satellite's internal equipment from external electrical charges. High-energy particles discharged by the sun rapidly lose strength as they pass through the multiple layers of a spacecraft's body or bus as well. There, they encounter a series of specially designed circuit dividers, individual compartments, and other unique structural elements that act as protection barriers.

The disruptive nature of solar weather impacts far more than satellite operations, and adversely affects terrestrial power and communications grids. For these and other reasons, a considerable amount of manpower and money has been devoted to monitoring the sun's activity, and more research into solar phenomena in general is planned in the future. Among other things, one benefit has been a steady improvement in our ability to rapidly detect and track these solar events using powerful observation and detection systems both on the ground and in space.

NASA, the U.S. National Oceanic and Atmospheric Administration (NOAA) and the DOD oversee much of this activity. For example, besides NASA's twin Solar Terrestrial Relations Observatory (STEREO) spacecraft, the Air Force Research Laboratory launched the Communication/Navigation Outage Forecasting System (C/NOFS) satellite several years ago to forecast the presence of ionospheric irregularities caused by the sun that adversely impact communication and navigation systems. Ground-based measurements also assist in space weather monitoring.

Satellites depend upon the sun, and satellite operators have steadily developed tools and techniques that allow them to ensure the operational integrity of all satellites in the face of all forms of solar weather. Thanks to proper planning, design and execution, solar events have had, to date, little impact on commercial satellite operations.

Conclusion

Dependable and ubiquitous satellite communication services are critical both to the global economy and to the national security of the United States. Because of the large capital investments required to design, build, launch and operate satellites, commercial operators have a vested interest in doing all they can to protect their spacecraft in orbit from the real threats posed by other objects in space, signal interference, solar weather, cyber-attack and intentional jamming. The U.S. Government also has billions of dollars invested in communications satellites and shares the industry's desire to protect its critical satellite communications capability. Because the Government relies so heavily on commercial satellite capacity, a spirit of cooperation is required to maintain the overall safety of the global satellite fleet, both commercial and Government owned.

Determining the orbit of objects near GEO is a complex task, particularly for uncooperative objects. As the population of objects in the GEO neighborhood grows, maintaining a secure and highly accurate catalog of all objects becomes increasingly important to mitigate risk of collision and for security of high value assets. The current Space Surveillance Network (SSN) capabilities for tracking these objects are subject to a number of constraints—particularly weather, scheduling, geographic diversity dispersion and overall capacity—which leave the current GEO catalog in significant need of improvement. Recent efforts, within SDA, to share owner-operator data provide clear proof of the value of collaboration.

The Space Data Association was established to allow all satellite operators to cooperate in tracking known objects in space. While NASA and NOAA have both joined in providing information to the SDA database, other U.S. Government agencies—most particularly, the U.S. Department of Defense—have not yet chosen to participate.

In its next evolution, the SDA will employ the resources and relationships it has developed to address the growing issue of radio frequency interference. While the user error that causes most RFI incidents will never be completely prevented, commercial operators now are deploying Carrier ID and developing other tools in place to quickly solve interference problems. However, this only applies to commercial satellites. Currently, there is little coordination between Government and industry when government-owned satellites are involved. This is an area where better cooperation could ensure that space assets are available to all users when they are needed.

In creating the SDA, the private sector has taken the first step towards a new paradigm for managing risk in space, but to be most effective, far more cooperation is needed by both commercial and government satellite operators worldwide.

Senator NELSON. And we will want to know, Mr. DalBello, how you "safe" your satellites when there is a solar flare.

Dr. Johnson-Freese?

STATEMENT OF DR. JOAN JOHNSON-FREESE, PROFESSOR, NATIONAL SECURITY AFFAIRS, U.S. NAVAL WAR COLLEGE

Dr. JOHNSON-FREESE. Thank you. It is my pleasure and honor to speak to the Committee today, just as it always is to speak to students and the public about space, which I do many times a year.

I am going to take a slightly different approach to assessing risks, impacts, and solutions for space threats by talking about the threat of the American public not understanding the importance of space.

From my course on space and security at Harvard to speaking to the public about the immediate importance of space activity in their lives, the number one comment I subsequently receive is, why don't we know this stuff?

Space, based on my interactions with the public, is not the final frontier; it is not the next frontier. In 1997, I coauthored a book calling it the dormant frontier, but it is not really that either. Quite the opposite, given the ambitious work being done on the International Space Station and other activities, though I contend that that work is largely unknown to the public.

Space is, however, for many Americans not living around a NASA center, a benignly neglected frontier. The problem with space and public support—support that translates into prioritized spending of their tax dollars—is that the public views space much as most people view their cars: they just want them to work. They don't care about the mechanics of a combustion engine or how to build or repair the car; they just want to drive the car.

Similarly with space, because of the resounding success of NASA and other organizations responsible for putting substantial space infrastructure into orbit, Americans—indeed, people all over the world—use their ATM cards, use GPS in their cars and boats, and rely on the Weather Channel to tell them whether they should carry an umbrella, totally oblivious to the role that space assets play in providing that information.

So, in that regard, the immediate benefits of space activity are not forgotten to most of the public; perhaps that knowledge was never known in the first place.

Space is associated largely with exploration, and so considered expendable during times of economic restraint, something that can be put off until later. But that premise is incorrect, even regarding exploration. Infrastructure isn't built or launched quickly, and, once in orbit, it must be maintained and updated.

Though, because they use it, the public is familiar with some satellite systems, like GPS, but less familiar with others. For example, the probes that study radiation belts and satellites that watch for the solar storms to protect other satellites in terrestrial electrical grids are essential. Unfortunately, however, the public generally views these space activities as little more than interesting science projects, if they know about them at all.

Yet, without them, Americans' lives would fundamentally change. Let me explain with a few brief examples. GPS is, with the Internet, one of only two global utilities. It facilitates, for example, having emergency response vehicles reach their destinations by the shortest routes, potentially saving lives; for transoceanic air travel to be safer and more efficient because planes can fly closer together; and if the new satellite-reliant air traffic control system is implemented, reduce jet fuel consumption by 1 million barrels annually, saving both money and the environment; and it saves the trucking industry an estimated $53 billion annually in fuel costs and better fleet management.

In addition to the economic benefits of space, which are vital to the national interest, there are also direct security implications. Politically, the recent meteorite that hit the Russian Urals with the force of an atomic bomb was a stark wake-up call regarding threats from space and certainly raised the importance of space surveillance and the communication of those threats.

Given the complex political state of the world, it is clearly imperative that government officials have accurate scientific data to distinguish between meteorites and missile attacks. Since miscalculation is a historic cause of war, we must be aware of what is going on in our solar system.

The military benefits of space are too lengthy to be catalogued in a short period of time, but suffice it to say that every letter in the acronym which basically describes military operations, C4ISR—command, control, communication, computers, intelligence, surveillance, and reconnaissance—is reliant on space assets. Space surveillance allows the military to have eyes and ears into locations otherwise inaccessible and on a 24/7 basis.

Geostrategically, America is hindered by our own success with Apollo. Especially in tough economic times, many Americans view space exploration with a been-there-done-that attitude. But for the rest of the world, space still represents the final frontier, the future, and, consequently, global leadership. Space capabilities add to U.S. prestige and soft power, which has spillover into U.S. influence in multiple policy areas.

Two final questions tie many of these issues together and hopefully illustrate why a vigorous space program must be maintained.

First, would China have conducted a high-altitude kinetic ASAT test in 2007, with the resultant debris threatening the sustainability of the space environment, if it had been a partner on the International Space Station?

Second, if called upon to deflect a meteorite threatening Earth, are the technologies in place to do so, and are the mechanisms in place so it could be done without a geostrategic nightmare?

In conclusion, America will stay ahead in space and, thus, capable of addressing these economic, political, military, and geostrategic risks and threats by staying active and staying ahead. Therefore, we must remind the American people and remember ourselves that space exploration and development is not expendable; it is in our strategic national interest.

[The prepared statement of Dr. Johnson-Freese follows:]

PREPARED STATEMENT OF DR. JOAN JOHNSON-FREESE,[1] PROFESSOR, NATIONAL SECURITY AFFAIRS, U.S. NAVAL WAR COLLEGE

It is my pleasure and honor to speak to this Committee today, just as it is my pleasure and honor to speak to students and the public about space many times each year. Based on that interaction with the public, I'm going to take a slightly different approach to assessing Risks, Impacts & Solutions for Space Threats by talking about the Threat of the American Public Not Understanding the Importance of Space. From my course on Space & Security[2] at Harvard Extension and Summer Schools, to speaking at the Mid-Coast Forum in Maine[3] in December 2012, and to the Blue Water Sailing Club earlier this month,[4] the number one comment I receive after talking about much the same material covered in my oral and written testimony today—the near-term importance of space to every American—is, "why don't we know this stuff?"

Space, in my opinion and based on my interactions with the public, is not the final frontier, and it is not the next frontier. In 1997 I co-authored a book calling it the dormant frontier[5] but that isn't really correct either—quite the opposite given the ambitious and effectual work being done on the International Space Station (ISS). I would posit though that few Americans are even aware of work being done on the ISS. If the general public is more than casually aware of the ISS at all, it is as a lame-duck program waiting to be de-orbited some time in the near future. Space, for many Americans not living around a NASA Center, is largely a benignly-neglected frontier.

The problem with space and public support, support as it translates into support for prioritized spending of their tax dollars, is that the public views space much as they view their cars. When they get into their cars, they just want it to run. They don't care about the mechanics of a combustion engine, or how to build or repair a car, they just want to drive it.

Much the same way with space, because of the resounding *success* of NASA and other organizations that have been responsible for putting space infrastructure into orbit, Americans—indeed people all over the world—use their ATM cards, use GPS in their cars and boats, and rely on the Weather Channel to tell them whether to wear a coat- totally oblivious to the role that space assets play in providing that information. I have had people ask me why the United States should invest in more weather satellites, when we already have the Weather Channel, unaware of the connection. So in that regard, the immediate benefits of space activity are forgotten to

[1] The views expressed here are the author's alone and do not represent the official position of the Department of the Navy, the Department of Defense or the U.S. Government. The author thanks the Naval War College EMC Chair for its support to participate in this hearing.

[2] After teaching this course in during the Fall Semester 2011 at Harvard Extension School—with a student population rich in diverse student backgrounds and perspectives—I wrote an article about student attitudes. See: "Guest Blog: Views on Space From an (Rare) Informed Public," *Space News,* January 5, 2011.

[3] http://www.midcoastforum.org/speakers/dr-joan-johnson-freese

[4] http://www.bluewatersc.org/news/27/BWSC-Visits-U-9S-Naval-War-College-Newport/

[5] Joan Johnson-Freese and Roger Handberg, *Space: The Dormant Frontier,* Praeger, 1997.

most of the public, or perhaps that knowledge was never known in the first place. Space is associated largely with exploration, and so considered "expendable" during times of economic restraint; something that is desirable and exciting to do, but can be put off until later.

But that premise is incorrect—even regarding exploration, as I will point out regarding the geostrategic importance of space. Infrastructure doesn't go up quickly—and once in orbit infrastructure must be maintained, and updated. And while some satellites—like GPS—are relatively easy for at least some of the public to understand because of they use it, other systems are less obvious. For example, the probes that study radiation belts and satellites that watch for solar storms to protect satellites and terrestrial electrical grids are essential.[6] Unfortunately though, the public generally views these activities as little more than interesting science projects, if they know about them at all. Yet, without them Americans' lives would fundamentally change. Let me to explain with a few brief examples.

GPS is—with the Internet—one of only two global utilities. Its usage allows us to, for example:

- use our ATM cards wherever we are in the world,
- to buy gasoline at the pump using a credit card,
- for emergency response vehicles to reach their destinations by the shortest possible route, potentially saving lives,
- for trans-oceanic air travel to be safer and more efficient because planes can fly closer together, and if the new satellite reliant air traffic control system is implemented, reduce jet fuel consumption by 1 million barrels annually—saving both money and the environment [7]
- and save the trucking industry an estimated $53B annually in fuel costs and better fleet management.[8]

In addition to the economic benefits of space, which are vital to the national interest, there are direct security implications of space.

The recent meteorite that hit in the Russian Urals with the force of an atomic bomb was a stark wake-up call regarding threats from space, and certainly raised awareness about the importance of space surveillance and the communication of threats. Press reports talked about the panic experienced by local residents, including that the world was coming to an end, and that one Russian official blamed the event on the United States.[9] Given the complex political state of the world, it is clearly imperative that government officials have accurate scientific data to distinguish between meteorites and missile attacks. Since miscalculation is a historic cause of war, we must be aware of our solar system.

The military benefits of space are too lengthy to be briefly catalogued, but suffice it to say that every letter in the acronym which basically describe military operations, C4ISR—Command, Control, Communication, Computers, Intelligence, Surveillance and Reconnaissance—is reliant on space assets. Space surveillance allows the military to have eyes and ears into locations otherwise inaccessible, on a 24/7 basis. It is also important to note that an estimated 80 percent of military communications is carried on commercial satellites,[10] making it imperative that the private space sector be kept healthy and viable. Without space-based assets, the United States would simply cease to be a superpower.

Geostrategically, America is hindered by our own success with Apollo. For many Americans, unfortunately, space exploration is viewed from a kind of been-there/done-that perspective. But for the rest of the world, space activity and space exploration still represents the Final Frontier, the Future, and, consequently, Global Leadership. Space capabilities add to U.S. prestige and soft power, which has spillover into U.S. influence in multiple other policy areas.

---

[6] Lisa Grossman, "NASA's tentacle spacecraft will probe solar storms," *New Scientist,* August 9, 2012. *http://www.newscientist.com/blogs/shortsharpscience/2012/08/tentacled-spacecraft-will-prob.html*

[7] Ann Schrader, "Air Traffic Control's Next Generation May Give Airlines' Fuel-Saving, Fliers a Lift, September 7, 2011, *The Denver Post, http://www.denverpost.com/business/ci_18840006*

[8] Eric Ogden, "Study: GPS Could Save the Trucking Industry $53 Billion," July 1, 2008. *http://www.informationweek.com/mobility/business/study-gps-could-save-trucking-industry-5/229211071*

[9] Associated Press, "About 1,100 injured as meteorite hits Russia with the force of an atomic bomb," February 15, 2011. *http://www.foxnews.com/science/2013/02/15/injuries-reported-after-meteorite-falls-in-russia-ural-mountains/*

[10] Stew Magnuson, "Military Space Communications Lack Direction," *Space Daily,* January 7, 2013. *http://www.spacedaily.com/reports/Military_Space_Communications_Lacks_Direction_999.html*

The U.S. is the global leader in space—one need only look at both the quantity and quality of spacecraft the U.S. has in orbit to verify that—but the U.S. has in some cases already lost the *perception* of being the leader in space, ceding that position to China.[11] The prospect of a U.S.-China Space Race has titillated the press and pundits for several years.[12,13] But the U.S. will cede space leadership to China not from the lack of scientific and technical potential or capacity in the United States, but due to the lack of political will to maintain it.[14]

Two final questions tie many of these issues together, and hopefully illustrate why a vigorous space program must be maintained.

First, would China have conducted a high-altitude kinetic ASAT test in 2007—with the resultant debris threatening the sustainability of the space environment—if it had been a partner on the ISS? This certainly question begs consideration of the best approach to dealing with Chinese space ambitions:[15] by cooperation, competition, or attempted isolation, the latter unlikely to be successful. The Baker Institute at Rice University recently recommended that China be included in the ISS partnership,[16] a recommendation with which I strongly concur. Ironically, however, there are individuals in the Chinese space and foreign policy communities who, though once interested in such a partnership, have lost interest because they feel the U.S. politics are too fickle for the U.S. to be a reliable partner, and consequently would hinder their own ambitious domestic space exploration efforts.

Second, if called upon to deflect a meteorite threatening Earth, are the technologies in place to do so, and are the mechanisms in place so it could it be done without it being a geostrategic nightmare? An International Space Code of Conduct[17] is currently being considered by on a global basis. Secretary of State Hillary Clinton endorsed the concept on behalf of the United States, and the Pentagon is on board as well.[18] This indicates recognition of a need for mechanisms or guidelines for international cooperation on space issues beyond national control, which encompasses nearly all space issues.

Space is increasingly described as a congested, contested and competitive environment. It is undoubtedly all of those in one way or another, which inherently means America must stay actively engaged. For that engagement to be effective though, will require the addition of another "C" to that list of descriptors—cooperative, as space and space activity is also inherently international, intercultural and interdisciplinary in nature.

In conclusion—America will only stay ahead in space and thus capable of addressing economic, political, military and geostrategic risks and threats, by staying active. We must remind the American people, and remember ourselves, that space exploration and space development is not expendable, it is in our strategic national self-interest.

Senator NELSON. Thank you to all of you.

---

[11] "China eyes lead in international space race," CBS News, July 11, 2011. *http://www.cbsnews.com/2100–205_162-20078365.html;* Clara Moscowitz, "US is losing in race to its own Moon," *Space.Com*, October 19, 2011. *http://www.space.com/13331-china-space-race-moon-ownership-bigelow-ispcs.html;*

[12] Peter Ritter, "The New Space Race: China vs the US," *Time*, February 13, 2008. *http://www.time.com/time/world/article/0,8599,1712812,00.html;* Clara Moscowitz, "US & China: Space Race or Cosmic Cooperation," *Space.Com*, September 27, 2011, , *http://www.space.com/13100-china-space-program-nasa-space-race.html;* Daryl Morini, "The Coming U.S.-China Space Race," *The Diplomat*, August 15, 2012. *http://thediplomat.com/china-power/a-u-s-china-space-race-in-the-offing/*

[13] Even in Russian press reports on the recent testimony of U.S. Director of National Intelligence James Clapper to the Senate Intelligence Committee, differing Russian views were given regarding whether China will try to threaten U.S. access to space, or whether threats are being hyped in the U.S. for budget purposes. "Will China Surpass the USA in Space?" *Voice of Russia*, March 13, 2013. *http://english.ruvr.ru/2013_03_13/Will-China-surpass-the-USA-in-space/*

[14] Joan Johnson-Freese, "Will China overtake America in Space?" *CNN*, June 20, 2012. *http://www.cnn.com/2012/06/20/opinion/freese-china-space*

[15] Joan Johnson-Freese, "A Long March Into Space," *Cairo Review*, February 10, 2013. *http://www.aucegypt.edu/gapp/cairoreview/Pages/articleDetails.aspx?aid=297*

[16] See: Marc Carreau, "Think Tank Recommends Role for China in ISS," *Aviation Week & Space Technology*, March 13, 2013. *http://www.aviationweek.com/Article.aspx?id=/article-xml/asd_03_13_2013_p05-01-558310.xml*

[17] Micah Zenko, "A Code of Conduct for Outer Space," Council on Foreign Relations, *http://www.cfr.org/space/code-conduct-outer-space/p26556;*

[18] Sydney Freedberg, "Why the Pentagon Wants an International Code of Conduct for Space," *AOL Defense*, March 20, 2012, *http://defense.aol.com/2012/03/22/safe-passage-why-the-pentagon-wants-an-international-code-of-c/*

And I want to welcome—we have a number of emergency responders that are in the audience, a new kind of threat that we are talking about that they may have to respond to.

And I suppose, in response to one of your questions, Dr. Johnson-Freese, that, as Senator Cruz had posed at the outset, maybe we ought to have Bruce Willis start doing another "Armageddon" movie to get everybody sensitized to the fact of how space could well play such a huge consequence in our lives if one of these asteroids starts coming toward us.

So let me turn to Senator Cruz.

Senator CRUZ. Well, thank you, Mr. Chairman. And there probably is no doubt that, actually, Hollywood has done more to focus attention on this issue than perhaps a thousand congressional hearings could do.

[Laughter.]

Senator CRUZ. Although I would not wish a thousand congressional hearings on anyone.

[Laughter.]

Senator CRUZ. You know, Dr. Green, Dr. Lu, I would like to go back to your testimony and get a little bit more in terms of the magnitude of the potential threat that near-Earth objects could present.

February 18 we had the meteor strike in Russia. Did we have any real warning of that strike before it occurred?

Dr. GREEN. Yes, it was February 15, it was really quite a special day because we actually had two events. The first one was a very close fly-by of a much larger asteroid that we call DA14. We had been watching that one for over a year, and we had calculated its orbit. We knew it was on a safe trajectory to pass by the Earth.

Indeed, the much smaller meteorite of 17 meters in size that struck Russia was not observed prior to its entry into the atmosphere. It was on a very difficult trajectory for us to be able to see from ground-based telescopes and came basically in the sunward direction.

So our telescopes operate from the ground in the evening, of course, on the night sky. One of the next major steps that has been now initiated by B612 is to provide a space-based asset that would plug that hole in a number of ways.

This is why our public-private partnership is extremely important for us to be able to continue to be able to help B612 with the Sentinel mission, to get up into space this decade, to then begin to observe many more of these objects that are on difficult orbits for us to see.

Senator CRUZ. If that same meteor, instead of striking a relatively rural area, had struck Manhattan, what would likely the consequence of that have been?

Dr. LU. This meteor exploded at altitude, 20 miles high or something like that, and, again, about 40 miles outside of the city. So, had it struck over a populated area, say, Washington or New York, it would have probably caused quite a few injuries, as it did over Chelyabinsk, but it wouldn't have taken out the city. It was too small to do that.

But, as you could see, even from a distance of 40 miles, which is a very long distance, it basically blew in windows and doors.

Every window in that city was busted. And remembering that the shockwave that caused that drops off very rapidly the further away you are from it, had that shockwave been much closer to a city, there would clearly have been a lot more injuries.

Senator CRUZ. Now, my understanding is we have identified nearly 1,000 near-Earth objects——

Dr. LU. Ten thousand.

Senator CRUZ. Well, nearly 1,000 that are a kilometer or more.

Dr. LU. Yes.

Dr. GREEN. Correct.

Senator CRUZ. What would the consequences be of an impact from an object of that size?

Dr. GREEN. Well, the large objects, the ones that are 1 kilometer and larger, are actually very bright. Starting 15 years ago from the 1998 congressional action, really these are the ones that we have been after. Indeed, we believe we have gotten well over 95 percent of them.

What we do when we attain that information is we calculate their orbits out to more than 100 years. It is from that database that it is clear to us that those currently, the ones that we know of, will not pose a hazard. However, we are constantly monitoring them, and we constantly see them through our surveys.

So, from that perspective, that speculation is just that: it is speculation. So what we want to continue to do is, of course, get into space, continue our program on the ground, and methodically, over a number of years, complete the survey so that we can see what the real threats are, rather than speculate.

Senator CRUZ. Now, Dr. Lu, you mentioned that you think our current knowledge is roughly 1/100th of what is out there?

Dr. LU. Yes, of those larger than the one that missed us on February 15, which is roughly the size of the one that struck in 1908.

Senator CRUZ. Yes.

Dr. LU. And so those are the ones that would be only large enough to take out a large city, for instance, not something that would kill off civilization or send us into the Dark Ages, but maybe only destroy New York City.

And of those asteroids, we know well less than 1 percent of those. So right now the amount of warning time that we are likely to get from one of those is zero.

Senator CRUZ. And let me ask a final question. What is our capacity if we discovered a sizable asteroid that was on a collision course? What is our capacity right now to do something to change that?

Dr. LU. If you find it early, decades in advance, which is what the goal of NASA is to do and goal of the B612 Foundation is to do, we have many options. Then you only need to change its trajectory by a very, very tiny amount.

Senator Nelson, you know from, you know, having flown in space, that when you are many orbits ahead of time, very tiny changes in your speed make big differences in the timing of where you are many orbits later. And that is exactly what you do. So, in real terms, if you change an asteroid's speed by something like a millimeter per second—you know, that is about the speed that an ant walks—and you do that 10 years or more decades before it is going to hit the Earth, you can make it miss the Earth.

So that means all you basically really need to do is either run into it with a small spacecraft; it is called a kinetic impactor. You can tow them gravitationally using a small spacecraft called a gravity tractor. For the very larger ones, the kilometer-sized ones, you can use a nuclear standoff explosion. These are all technologies that we believe we know how to do.

The key is, if you don't know where they are, there is nothing you can do. If you have less than a few years' notice, right now we have no options.

Dr. GREEN. There is another aspect of this that I want to mention, and that is, these objects are very heterogeneous. They can be rubble piles. Their composition is quite different. Some have iron, some don't. Some are carbonaceous chondrites, stonies. Consequently, it is important to know what you are up against.

In fact, this particular decade is a great decade for us to be able to do some of the research necessary that will contribute to potential mitigation concepts into the future.

One mission is OSIRIS-REx. It is an asteroid-sample-return mission. It is going to an asteroid that is called RQ36, which is a potentially hazardous asteroid in more than a 150 or so years. But it gives us an opportunity to get up close to it, grab a sample. We will orbit it for more than 500 days. We will understand how the solar wind and the light from the Sun potentially moves the object.

So this kind of study is essential for us to be able to really determine the potential mitigation strategies that we would use in, potentially, future missions that we may have to pull off.

Senator NELSON. So, right now, until Dr. Lu gets his satellite up there in 5 years, we are just hoping that we can identify and then correctly calculate the trajectory that one of these asteroids would have.

And assuming that we found one before you got your satellite up, let's go back to Senator Cruz's question: What would an asteroid that is a kilometer in diameter, what would it do if it hit the Earth?

Dr. LU. That is likely to end human civilization.

Senator NELSON. So that was typical of maybe what hit at the time of the dinosaur age?

Dr. LU. No, that asteroid was yet much larger, another factor of 10 larger, 10 in diameter larger, which makes it 1,000 times more massive. And that led to the extinction of essentially 90 percent of all species alive at the time. And those are quite rare. And so we do not know of anything that large that is on an impact trajectory.

Senator NELSON. Does the fact, as we prepare to go to Mars and the stated goal of rendezvousing and landing and returning with a human crew from an asteroid, does that in any way help us perfect our ability to avoid this kind of catastrophe?

Dr. LU. I think, obviously, there is great science you can do. I think likely the deflection mission that we have to mount someday—and we will have to someday; we know that, someday—is likely to be done robotically, just because the distances are quite large from the Earth.

But there is a connection between the two, in that, again, for the human missions to asteroids, you still need to find them. We do not currently have a set of good targets to run human missions to asteroids. So the same data set which allows us to know if something is going to hit Earth gives us targets for exploration.

Dr. GREEN. Yes, in fact, visiting an asteroid by astronauts, for instance, is another one of those steps in terms of understanding much more about their characteristics. But the ability to do that is an enabling one. It is one of those where you trek outside of low-Earth orbit. You have a destination. You then go through a variety of processes and procedures that you would have to perfect on even longer voyages if you were to go to Mars.

So there are different objectives with respect to that. But, indeed, learning much more about our heterogeneous asteroid environment is incredibly important.

If I can just build on what Ed has mentioned, 50 years ago, planetary scientists believed that all the craters that are on the Moon were volcanic. We didn't know that they were of impact. It really took many years for us to be able to realize that they were created by impacts.

Then, soon after that, we began to notice the impacts here on the Earth. The impacts, of course, led us to the Chicxulub Crater, which indeed now is believed to be one that has created the extinction of the dinosaurs. In the 4–1/2 billion years of this planet, there have been five extinction events, for which that one, we do believe, has been the one done by a near-Earth object.

So over the last 50 years, we have learned an enormous amount about a brand new field, things that we needed to be aware of. And we have been putting in place a methodical program of observation by starting with the ground, by building on international partnerships, and also now with our public-private partnership with B612.

Senator NELSON. Well, until Dr. Lu gets his satellite up there, I hope you are paying your trajectory specialists well, Dr. Green.

[Laughter.]

Dr. GREEN. Indeed we are.

Senator NELSON. Because they have to be right on the ones that we know to make sure we know their trajectory.

Tell us, Dr. Lu, and don't worry, I am getting to the other two of you. Tell us about how we are going to nudge this thing. And what is the potential cost? And are we getting the international community understanding that they have to participate in this with us?

Dr. LU. Well, I think the—again, I mentioned the way you would likely do it, in most cases, is to simply run into it with a small spacecraft. Again, you would just need to change its velocity by a very, very, very tiny amount.

Following that, you will probably want to hover a gravity tractor near it to verify that you have changed it the way you think you did and to make any fine-scale corrections on it, something like a vernier burn on the Space Shuttle. You would do that with a gravity tractor.

The cost of that, such a mission, I would believe it would probably be in the range of a billion dollars or more; to do a couple of missions like that, probably a couple of billion. But I think you would have to compare it against the potential losses of a multi-megaton impact.

Also, in terms of the ability for the United States to show leadership, clearly, such a thing would be led by the country that has the greatest technological capability, which is the United States and NASA in particular. I think the world will be involved in both the decisionmaking process and the actual implementation of such a thing.

And I just, again, want to point out that there is a 30 percent chance that there is a 5-megaton or so impact that is going to happen in a random location on this planet this century. So this is not hypothetical.

Senator NELSON. By the way, why did this one in Russia explode at 20,000 feet?

Dr. LU. Tremendous deceleration. You know, when you are moving that fast—that thing was moving about 12 miles per second, and, at that speed, even hitting the air is like hitting concrete. And once it begins to come apart, it really rapidly comes apart, and it basically exploded.

Senator NELSON. And back to Senator Cruz's question. Had that occurred 20,000 feet over New York City, other than blowing out windows, what would have happened?

Dr. LU. There would have been, obviously, many casualties. This one is a little bit harder to say.

The one over Tunguska, the one that was slightly larger and the ones that I have been talking about, had that happened over a city, say New York City, we would have 7 million casualties, at least. Whatever the population in New York City is, they would be gone.

Senator NELSON. Really?

Dr. LU. Yes. The area of destruction of that impact was about 800 square miles. So there is nothing standing in an 800-square-mile-area forest where that impact occurred.

Dr. GREEN. Just to be clear, Ed is talking about the event that occurred in 1908 called the Tunguska event.

Dr. LU. Yes, not Chelyabinsk.

Dr. GREEN. That was a much larger meteorite.

Senator NELSON. What was the size of it?

Dr. GREEN. We estimate it was about 50 meters.

Dr. LU. Yes.

Senator NELSON. Fifty meters.

Dr. LU. Yes.

Dr. GREEN. Yes.

Dr. LU. A little bit larger than this room, about twice the size of this room.

Dr. GREEN. Now, a 50-meter impact, we believe occurs on the Earth every few hundred years.

So it is a rare event.

Senator NELSON. The one over Russia was about the size of this room.

Dr. LU. Yes.

Dr. GREEN. Seventeen meters.

Senator NELSON. Did the one in 1908, did that hit the Earth, or did it explode in the air like this most recent one?

Dr. LU. It exploded in the air.

Senator NELSON. I see. And the one that just entered the atmosphere over Russia, other than causing those windows —how many miles away from that little city in Siberia was it?

Dr. LU. Over 40 miles away.

Senator NELSON. Wow. Did it flatten the forest?

Dr. LU. Not really, no, because it exploded at a higher altitude than the one in 1908. So it was a huge shockwave, which doesn't blow down trees but it certainly knocks in windows, as we saw.

Senator NELSON. And if that one had exploded over New York at 20,000 feet—you were making reference to the 1908 one—what would that have done to Manhattan?

Dr. LU. If Chelyabinsk had exploded over New York City, we would have a lot more because it would be a lot closer than 40 miles, obviously—then we would have a lot more than broken windows, that is for sure.

Senator NELSON. Mr. DalBello, you have all these Intelsat satellites out there. What do you do to "safe" them when you have a solar explosion?

Mr. DALBELLO. Well, that is an interesting question. The truth is that these satellites are on and they are operating. So a typical satellite today is probably about 70 percent, actually, 70 percent full, so operating near its full capacity. And it is transmitting pretty much 24 hours a day. So there is really no opportunity to turn them off without significant disruption to the service they are providing. That could be banking or media or the transmission of other important information or the flights of UAVs or communications with the military. So there is no real off switch on our satellites today. So they are on.

So for us, the question is—and this gets back to the issue I was discussing earlier and the fascinating work that Dr. Green and Dr. Lu are doing. The question is always, how do you translate solar storm information into actionable warnings? When it is appropriate to translate information into practical warnings for industry?

So in the solar example, a little bit more—I guess a little bit more practical than some of the discussion on the asteroids, but the same question applies. How do we know what level of solar event will translate into a real impact on the satellites that we are flying?

Intelsat buys its satellites from the major manufacturers—Boeing, Loral, Orbital—primarily here in the U.S. These manufacturers are, of course, trying to build their satellites to operate in any environment. And the good news is that most satellites live well beyond their 15-year design life, so the manufacturers are doing a great job.

So the question for us is, at what point would an event be so extraordinary that we would say, "Okay, it is dangerous to do something with the satellite now; we are not going to load code today, or we are not going to try to communicate with the satellite today, or we would maybe put the solar panels in a different orientation." These are issues that just aren't really very clear right now. So we would look to a collaborative effort with the manufacturers and the government to provide advice on this.

Now, a similar thing happened in space debris. Several decades ago, NASA got very interested in understanding the debris environment. That led to the situation where we have today, where we actually get warnings from the Defense Department—they are called conjunction summary messages—we actually get warnings in advance if DOD thinks that there might be a collision between two space objects.

But it took us a long time to connect the basic information that we needed to gather to have the knowledge to the point where we felt comfortable warning people that they might want to take specific action, in this case moving a satellite. So I think the same situation will apply with solar. As we get more sophisticated in understanding solar storm impacts, then we will be able to translate those into specific actions. But today that knowledge just doesn't exist.

Senator NELSON. Senator Cruz?

Senator CRUZ. You know, going back to the discussion about potential meteor impacts, I would assume for an impact that the most likely place for an impact would be in water, given the percentage of the Earth that is covered by water.

What is the relative severity of an impact on land versus an impact on water and the consequences that would flow from one versus the other?

Dr. LU. It depends on the size of the asteroid. To cause a tsunami, it actually has to be large enough to strike the surface. And, as we saw, the smaller asteroids do not strike the surface. They explode in midair, which is—so the smaller asteroids, when they are over land, that can be more catastrophic. For the larger asteroids, when they strike the water, that could potentially be more catastrophic. And the way to think about is, essentially, these larger ones create a crater in the water, which then fills in.

So I don't know if you have ever been to Meteor Crater in Arizona. It was created by a small iron asteroid, and it is about 700 feet deep. It is about a kilometer across. And so, had that hit the water, you would have a 700-foot-deep crater in the water, momentarily, which then fills in. So you know the size of the wave; it would be about double that. You know, if you think of a rock dropping into a pond, right, you get a wave that is about twice the height of that. So you would get a 1,000-foot wave or so, something like that, coming off of something like that, which then drops off the further away you get from it.

So that is one of the great worries, especially with a couple hundred meter asteroids, the football stadium size asteroids. They are likely to hit in the ocean, and you are likely to have tsunamis. And, you know, as we saw in Fukushima, a tsunami can cause great damage and can affect—you know, that tsunami, which was not very large, historically speaking, had a noticeable effect upon world GDP because of damage that it did in one prefecture north of Tokyo.

Senator CRUZ. Do we have indication of near-Earth objects striking the water in the past and producing tsunamis?

Dr. LU. It must have happened before. It is a little bit difficult to distinguish a tsunami that was caused by an earthquake from one that was caused by an asteroid. At least, I mean, in the past, you know, in the historical record.

Senator CRUZ. And using the example you used about the size of asteroid that struck Arizona, if—and what was the size that we expect that was?

Dr. LU. The estimates of that are it is in the range of about 30 meters or so, so smaller than Tunguska but larger than Chelyabinsk.

Senator CRUZ. OK. But that would likely not be large enough—I guess there it did strike, so——

Dr. LU. There it did because——

Senator CRUZ.—it didn't explode in the atmosphere.

Dr. LU.—that one happened to be made out of iron, so it was a lot tougher than the stony ones, which can explode at higher altitudes. So even though that was smaller, it hit the ground.

It is worth a visit, by the way.

Senator CRUZ. I have not been, but I will have to add it to the list of places to take my daughters, and hopefully it—well, what are the odds of the same spot getting struck twice?

[Laughter.]

Dr. LU. Pretty low.

By the way, there is an interesting impact site in Texas, just— it is about 100 miles or so to the east of El Paso. We used to see it in our T–38s flying back from El Paso all the time. John Young and I used to like to fly over it. John loves impact craters.

Senator CRUZ. So, using that example, you said that if a similar impact were to occur in water, we would see a 1,000 foot tsunami. What kind of distance would that be expected to travel where it would maintain——

Dr. LU. It depends greatly upon where. You know, the shape of the ocean bottom, the depth of the water, and so on. So I don't have a good answer for you.

Also, the characteristics of that tsunami are going to be a little bit different than earthquake-caused ones, which we understand much better because those are sort of done by a line in a fault, whereas this is more of a point, more like dropping a pebble into a bathtub.

And so the answer is, basically, it depends.

Senator CRUZ. You also testified about your estimates of the probability of a 5-megaton incident or a 100-megaton incident, which, if I remember right, was 30 percent and 1 percent, respectively?

Dr. LU. Yes. In the next century, in this century, yes.

Senator CRUZ. Could you provide a little bit of the data that go into those probability estimates?

Dr. LU. Yes. Yes. In fact, that is—there is not a lot of scientific disagreement about that. This was documented well in the National Academies' report of 2010 called "Defending Planet Earth," and that data comes from NASA. And there isn't a lot of dispute about that.

And just for your information, where it comes from is from three different sources. You can count craters on the Moon. You can look at your asteroid surveys with telescopes. And, finally, there are DOD assets that look down upon the Earth for rocket launches and nuclear weapons tests, but mostly what they see is asteroid impacts that exploded in the atmosphere. And much of that data has been declassified.

And those three independent means of measuring the numbers of asteroids agree with each other, and that is why we have fairly high confidence in it.

Senator CRUZ. Let me ask a final question, which is, in your professional judgment, and I would ask this to anyone at the panel who would wish to answer it, what else should we be doing to assess the threats that could seriously jeopardize human life and to be in a position to prevent those threats?

Dr. LU. Well, from my standpoint, on this particular problem what we need to do is an extensive survey of the objects in our own solar system. We know the locations and trajectories of the million nearest stars because our telescopes can look away from the Sun. We do not know the locations and trajectories of the million nearest asteroids, and yet those things hit the Earth sometimes. And I think that is a big, gaping hole, and our organization is working with NASA to fill that hole.

And I think that is tremendously exciting not just simply from the preventing-death standpoint, which is obviously a wonderful thing, but from the exploration aspect of it and the inspirational aspect of it. Because, again, I think a demonstration that humanity can work together to go out there and do something incredible—changing the Solar System to prevent our planet from being hit is an incredible demonstration of science, technology, mathematics, astronomy, and all the things that make our country great.

Senator CRUZ. Are there concrete steps that, in your judgment, would be prudent, beyond launching the Sentinel satellite that you are working on, that we should be taking in order to be aware of the potential risks?

Dr. LU. Well, in the meantime, obviously NASA is running its own searches, and those need to be supported.

They also just began supporting a telescope system called ATLAS, which is just at a couple million dollars. It is a fairly inexpensive system. What it is basically going to do is look for asteroids just before they hit, not to be able to prevent their impact, but potentially to be able to evacuate an area. So you might get a day or 2 or 3 notice before something hits. And the sole purpose there would be to simply, you know, get out of the way, to the extent you can. And I think that is a great program. And it involves lots of small observatories around the world and students working on it, and I think that is a wonderful thing.

Dr. GREEN. Now, from my perspective, let me build a little bit on what Ed said. That is, indeed, we have methodically had an observation program in place from the ground. We have also used other space assets. One is the WISE mission, which was an astrophysics infrared mission, that had a particular orbit that made it appealing to be used to look for near-Earth objects. That was quite successful.

That really was the proof of concept for that next step, and that next step is, indeed, as Ed mentions, a survey, an infrared survey. So we are working with Ed and taking that step.

In addition, we have a really aggressive program to uncover much more of the characteristics of these bodies. We need to know those. As Ed mentioned, their compositions are very different. Some are iron, and they pack more of a wallop. So we need to go out and we need to understand other aspects of that that would feed into a mitigation strategy.

That is one of the reasons why OSIRIS-REx, the next mission that we are going to do from a science research point of view, is also going to help us inform a potential mitigation strategy.

So we have in place a methodical program that we need to continue to work on and execute over this next decade.

Senator CRUZ. And what is our ability in terms of a near-Earth object that is not that near to determine the composition of that particular object?

Dr. GREEN. Well, it can be done in a variety of ways. We have opportunities to hit it with radar. Radar is incredibly important. We have facilities in Puerto Rico that we use, National Science Foundation, that we work with on that, in addition to our Goldstone radars. This enables us to get ideas about surface composition in a small way. We also from the ground make spectral observations. Indeed, from space, even several important points in the infrared will tell us a lot about its composition. So, from a scientific point of view, we are doing a lot in that particular area, and that is helping us classify these and understanding their origin.

Senator NELSON. Dr. Johnson-Freese, you are constantly trying to get people to understand the relevance of our space program. There would be nothing like focusing the mind than survival with one of these things heading toward us.

What about these and other threats, to protect life on Earth and protecting our astronauts, with all of the collisions that we find going on out there with space debris? How are we going to get the human space exploration to be conveyed to the American public just what is at stake here?

Dr. JOHNSON-FREESE. I think that goes back to Bruce Willis and "Armageddon." And I am all for Bruce Willis testifying, don't get me wrong, but after that movie came out I was part of a project called "Armageddon: Fact and Fiction." And what that movie did was basically convince the American public that if anything bad happened, people would get in the shuttle and go fix it. It was myth, it was not reality.

I think what we need to do is get far more of the information of the kind that Dr. Green and Dr. Lu have conveyed today to the American public. When I talk about it in class, when I talk about NEOs in class, the overwhelming response of my students is, "I don't know anybody who ever died from a NEO. I know people who have died of cancer, traffic accidents, but nobody I know has died from a meteorite." And this gets into, again, the idea that we need to convey more of the fact and separate it from the fiction that the movie industry has really convinced much of the American people, that we are all over it, we can take care of this.

So I think there needs to be—exploration and vision and inspiration is wonderful and has to be a component of our space program. But I think we need to get much better and much more aggressive at conveying the risks and the benefits and the self-interest in not just continuing but expanding the space program.

Senator NELSON. Well, the American people are certainly appreciative of the conveniences——

Dr. JOHNSON-FREESE. Absolutely.

Senator NELSON.—that they have every day, but do not know the connection, as Mr. DalBello has said, that all of these conveniences happen to be space-based, one way or another.

Now, you take—that is why I ask about solar explosions, which is a nuclear explosion on the surface of the Sun. It emits radiation. Unless satellites are safe or they are within the magnetic sphere surrounding the Earth that would repel this radiation, there is a possibility they are going to be knocked out.

Let's take another scenario. What about a rogue country like Korea or Iran, if they get a nuclear weapon—which we certainly hope and it is the United States' intention that they don't. But what if they put it on one of their rockets in North Korea, sent it up at altitude, and exploded a nuclear weapon? That would have some rather serious consequences, wouldn't it, Mr. DalBello?

Mr. DALBELLO. It would be a very, very bad day for satellites, yes.

Senator NELSON. Explain that.

Mr. DALBELLO. The radiation environment——

Senator NELSON. Explain that so our audience will understand it.

Mr. DALBELLO. Actually, we went back—when there was upper-altitude testing previously, we did have evidence that the way it energizes the orbits also interferes with the electronics of the satellite. And depending on where it is in altitude and which portion of the orbit you are in—obviously, satellites at geostationary orbit, which are 23,000 miles away, are a little bit safer. Satellites in lower-Earth orbit would be saturated soon and would probably die, because the electronics would be saturated by the energy released from the nuclear explosion.

So a high-altitude/near-space explosion could have a very catastrophic effect on many of the satellites we rely on for weather, early warning, imagery, and other very critical functions.

Senator NELSON. Dr. Lu, what would that do to our astronauts on the Space Station?

Dr. LU. It clearly wouldn't be good. You know, I have had the experience of being told to take shelter on board the International Space Station because of a large solar flare. That happened in 2003, and it has happened a few times since. But, you know, these levels of radiation could be much higher.

Senator NELSON. Senator Cruz?

Senator CRUZ. I am good.

Senator NELSON. Staff?

Mr. DalBello, your company has over 50 satellites in orbit.

Mr. DALBELLO. Correct, more than 50 that we own. We fly over 70 satellites total because we also fly satellites for other operators.

Senator NELSON. How do you build the risk to these satellites into your business model?

Mr. DALBELLO. Well, planning for a fleet of that size means that you are always doing several things. First of all, you are always building new satellites. You are always planning the launch of those satellites. And, again, this is well beyond the topic of discussion today, but launch is still a problematic area for us. We wish the industry were much more robust and reliable. We welcome the entry of SpaceX and other new entrants into the marketplace. But it is still a challenging and expensive component.

So you are obviously building satellites, you are preparing to launch those satellites, and you are managing your fleet by moving satellites around in orbit in a way that, in the past, actually, we really didn't do.

Typically, in the past, we would put a satellite in one place, and it lived and died in that orbital location. Now it is much more dynamic, and we are constantly grooming the fleets. We move satellites to meet demands. For example, when the war in Iraq started, we actually moved two entire satellites to the Middle East to accommodate the increase of traffic in the region.

So it is a really dynamic equation, and so we do occasionally get anomalies. And when something bad happens—unfortunately, we lost a brand new satellite a few weeks ago when the launch vehicle failed, the sea-launch vehicle failed. And that satellite was supposed to supply a lot of services. One was a military component in the UHF band, but it was also supplying a lot of television service to Latin America. And so we have to scramble to try to find replacement capacity. Sometimes we can find that in our own fleet, and sometimes we have to go to our colleagues and competitors to get that kind of capability.

So the same thing applies more generally to any effect which perturbs the fleet. You know, there is not a whole lot of excess capacity in the sky, so it requires a creative and constant maintenance of that capacity and understanding where our requirements are.

Senator NELSON. The vote has started, so we will wrap up in 5 minutes.

Space debris is really a problem. And I was struck several years ago, the deafening silence, lack of criticism of the Chinese when they launched their ASAT and blew up a satellite, adding tens of thousands of pieces of space debris that are up there that just add to the problem. And the problem, even if we didn't launch another rocket on Planet Earth, you would have a real problem of space debris up there for some period of time.

So, Dr. Green, what space debris removal technology is NASA considering?

Dr. GREEN. That is a good question. I know there is a variety of studies that they are working on, and I will have to get back to you with much more of the details of those studies.

Senator NELSON. Dr. Johnson-Freese, what are the legal and national security barriers to space debris removal?

Dr. JOHNSON-FREESE. One of the big barriers is that there are no salvage laws in space. So if the United States were to start an initiative today to clean up all the debris, we don't own it, we can't just go and get it.

And you decide, are you going to get the little pieces? That would determine if you are going to use some of the techniques like foam to catch it. Or do you go for the big ones? And from a legal perspective, if I went after a big piece of junk and grappled it and it broke apart, am I then legally liable for the damage caused to Rich DalBello's satellites?

So there are many legal issues to be considered. And then the political and geostrategic—if, for example, you are using lasers, well, I am certain if the United States decided to start using lasers to de-orbit large pieces of debris, that would make other countries of the world very nervous, just as it would legitimately make the United States very nervous if other countries were to do the same.

So I think there is a host of not just technical problems but legal and political problems. But debris and the NEO issue, I think, provide opportunities for international collaboration—require international collaboration and for the U.S. to take a real leadership role.

Mr. DALBELLO. Senator, can I make a brief comment about that?

First of all, I completely agree with what Dr. Freese just said, but we are very interested in—there is a next generation of technologies that is not very far away. We know that DARPA is looking at this, we know that NASA is looking at this. We know that there are a couple of private sector opportunities that people have come to us with proposals.

We are very interested in the ability to do more in space robotically. That is because—for example, we had a satellite that went up a couple years ago. One of the antennas didn't open, which meant that half of the revenue for the life of that satellite was lost. Had you been able to go up, grab that satellite, and just tweak that antenna a little bit, you probably could have saved the entire mission.

So we think the technology is there today, and perhaps this is a good opportunity for another hearing. We know that DARPA has some forward-leaning programs that they are working on. And it is a technology that we in the private sector support fully and want to participate with government on because it is a valuable thing.

That technology would also allow you, at a minimum, to remove very large pieces of debris from the most useful orbits.

Senator NELSON. Senator Cruz, anything?

Senator CRUZ. I just want to thank all four of you for what I think has been a very interesting and productive hearing.

And I thank the chairman, as well.

Senator NELSON. Indeed, it has been most enlightening. We thank you for your expertise and your testimony.

Have a great day. The meeting is adjourned.

[Whereupon, at 11:18 a.m., the hearing was adjourned.]

APPENDIX

RESPONSE TO WRITTEN QUESTION SUBMITTED BY HON. BILL NELSON TO
DR. EDWARD T. LU

Question. Dr. Lu, once Sentinel detects the many thousands of 140 meter asteroids and completes its mission, how do we sustain our detection efforts so that we don't lose track of these objects?

Answer. The orbits of the asteroids detected by Sentinel will be valid for approximately 100 years. That means that the complete asteroid survey does not need to be repeated for quite some time. However, it is important to maintain our capability to do targeted follow up observations for particular asteroids of interest that may pass uncomfortably close to Earth.

RESPONSE TO WRITTEN QUESTION SUBMITTED BY HON. BILL NELSON TO
RICHARD DALBELLO

Question. We're currently tracking over 21,000 pieces of space debris larger than 10 centimeters in diameter, but there may be many thousands of pieces even smaller we can't track accurately. Mr. DalBello, if NASA and Air Force tracking of space debris weren't available to companies like yours today, how would you protect your satellites?

Answer. To avoid collisions in the increasingly crowded orbital arcs, companies operating satellites have, for years, informally shared position and orbit data amongst themselves. Over time, operators have sought better information sources and formed cooperative structures—like the Space Data Association Limited (SDA)—to ensure that their operations were as safe as possible. Over this same period, the U.S. Government has slowly developed mechanisms for working with the commercial operators. [See Brief Timeline below]. In the absence of U.S. Government information, satellite operators would look for additional data sources from Europe, Russia, and private sources, however, none of these today are as comprehensive as the U.S. Government data. Without U.S. Government cooperation, all space operators—including the U.S. Government—would face increased risk and the potential degradation of the space operating environment. In particular, without data on debris, no satellite operators would be able to screen their satellites for potential collisions. Given the limitations of the current legacy software and processes currently used to provide that service and DOD's higher priorities for military space operations over supporting commercial or civil space operations, the potential exists for threats to be missed. The consequences of a collision—particularly in GEO—could be catastrophic. Not only would billions of dollars of commercial investment be at risk, but so would more than 90 percent of all military communications worldwide.

Brief Timeline of Information Sharing

It is important to note that accurate space tracking is a non-trivial task that has changed considerably over time. The following is a brief overview of the history of U.S. Government/industry data sharing:

- In the early-1980s, NASA provided orbital data (called two-line element sets or TLEs) to private individuals via mail. In 1985, one of those individuals began distributing that data electronically, first via a dial-up bulletin board and then via the Internet. This service became known as CelesTrak. This service still exists today and is operated by AGI.

- In the late 1990s, NASA began an informal program, run by their Orbital Information Group, to share this information on space objects gathered by the U.S. Strategic Command (USSTRATCOM) using radars and optical sensors. In 2005, the NASA dissemination was terminated and transferred to Air Force Space Command,.

- In 2004, Congress authorized DOD to establish a pilot program called the Commercial and Foreign Entities (CFE) Program. Through the CFE program, the U.S. Government began providing the same orbital data directly to commercial operators.

- Recently, USSTRATCOM acknowledged that the TLE data was not precise enough to screen satellites for close approaches and developed a procedure for providing commercial operators with additional information in the form of conjunction summary messages (CSMs). The CSMs are provided to operators whose satellites have been identified as closely approaching another space object. These CSMs contain state vector (position and velocity) and covariance (uncertainty) information computed from other data.

- Even with government information, informal information sharing between operators is problematic. Satellite operators use multiple standards to represent the position of a satellite in orbit or an object in space. Many different types of software are used to track and maneuver satellites and the data is stored in a variety of formats. So, even operators who wish to share data can't rely on a single, agreed upon protocol for sharing information. As a result, operators sharing information must maintain redundant file transfer protocols and tools to convert and reformat data so that it is consistent with their own software systems to compute close approaches. While some operators use third party software for predicting close approaches, others write their own software tools. As the number of satellite operators increases, the problem of maintaining space situational awareness grows more complex. And the smallest operators may not be able to afford, or have the technicians, to participate in the data sharing process.

- In 2009, the world's leading commercial satellite operators formed the Space Data Association Limited (SDA) to formalize the process of exchanging information and to deal with the overall data compatibility problem. Clearly, the best path to minimize risk in space is for all operators to share what they know about the movement and position of their own satellites, including any planed maneuvers in advance of execution, in a way that all other companies can use.

————

RESPONSE TO WRITTEN QUESTIONS SUBMITTED BY HON. AMY KLOBUCHAR TO DR. JOAN JOHNSON-FREESE

Question 1. In Minnesota, GPS technology has transformed many industries, making them more efficient. From farmers' tractors that steer themselves, to safer and quicker road construction, to navigating ships on the Great Lakes, high-precision GPS has made our economy more competitive. What can be done to make our network of GPS satellites more resilient in the face of solar storms?

Answer. There are two actions which address your question. First, the best protection is prevention. That means funding the satellites which observe the solar storms and allow scientists to predict when potentially dangerous flares will occur so that actions can be taken to protect the GPS satellites from damage. Second, the constellation of GPS satellites themselves will need upgrading that incorporates the latest technologies for protection, and GPS "spares" must be kept on orbit as well for inevitable instances when needed.

Question 2. Additionally, though many utilities are prepared to respond to downed power lines in the face of severe weather, what steps can we take to prepare the electric grid for the widespread damage to key infrastructure that could occur after a massive solar storm?

Answer. Again, prevention through prediction of the solar storms is a solid first step, so that power companies can take necessary actions—potentially including shutting down certain facilities—to avoid damage.

Question 3. You mentioned in your testimony that Americans aren't sufficiently aware of the work being done by NASA and other organizations, that the American public is often unaware of the connection of space work to many of the things we rely on a daily basis like GPS, credit cards, and weather satellites. Americans, you say, largely equate space to exploration. What are your thoughts as to why there's a disconnect between what the American public *thinks* space programs do and what they *actually* do? What can we do to better connect the dots?

Answer. In many instances, "space" is taught as a history lesson as part of a science curriculum in schools, and often in an outdated way. Yet space infrastructure is much like the railroad infrastructure or the computer industry—it provides services. So students must be exposed to and taught about space differently. Additionally, too often media people from whom much of the public gets its information are ill-informed, and consequently pass along wrong or sensationalist information.

I heard a local morning radio broadcast recently berating NASA over the new accelerated asteroid mission—wrong in their information and analysis, and passing it along to a wide audience. My advice to do better: education, education, education (perhaps biased because I'm a teacher).

Question 4. You've laid out concrete examples of important functions that space programs serve, but in a time of belt-tightening across all Federal agencies, how do we strike the right balance of cutting programs when necessary, yet ensuring these programs have the capabilities to perform?

Answer. I have stated repeatedly that more consultation between policy people and scientists/engineers is needed regarding what programs have the potential to reach fruition and be useful, and those which are programs of little value and with large science/technology question marks. Most of those are in the Defense Department. For NASA, programs should expand capabilities, rather than repeating past successes. Hence, I strongly support the accelerated asteroid mission: it provides a role for both robotics and human spaceflight, will return valuable information to protect Earth from an event that is not a question of "if" but "when," and will offer the U.S. the potential to take a global leadership role in space, again. That is important for both scientific and geostrategic reasons.

○

www.ingramcontent.com/pod-product-compliance
Lightning Source LLC
Chambersburg PA
CBHW081232170526
45165CB00009B/3046